MATHEMATICAL PREPARATION FOR GENERAL CHEMISTRY, second edition

WILLIAM L. MASTERTON
University of Connecticut, Storrs, Connecticut

EMIL J. SLOWINSKI
Chairman, Department of Chemistry
Macalester College, St. Paul, Minnesota

Saunders College Publishing
Harcourt Brace Jovanovich College Publishers
Fort Worth Philadelphia San Diego
New York Orlando Austin San Antonio
Toronto Montreal London Sydney Tokyo

This book was set in Times Roman by Intergraphic Technology, Inc.
The editors were John Vondeling, Maryanne Miller, and Kate Mason.
The art & design director was Richard L. Moore.
The text design was done by Nancy E. J. Grossman.
The cover design was done by Larry Didona.
The new artwork was drawn by Tom Mallon.
The production manager was Tom O'Connor.
This book was printed by Vail Ballou.

Mathematical Preparation for General Chemistry ISBN 0-03-060119-3

Library of Congress catalog card number 81-53068

234 090 98765432

Harcourt Brace Jovanovich, Inc.
The Dryden Press
Saunders College Publishing

Preface

The primary purpose of this book is to help students who are weak in the basic techniques of mathematics used in general chemistry. Some of these students appear to be victims of the "new math," until recently so popular in the high schools. Others are returning to college after some years and find that their mathematical skills have atrophied with disuse. We hope that by studying this book and working the problems provided, such students will improve the mathematical proficiency that is so vital to success in chemistry.

Chapters 3, 4, and 5 are devoted to the concepts of percent, exponential numbers, and logarithms, in that order. Chapters 2 and 6 cover two important areas where simple mathematics impinges directly upon general chemistry: unit conversions and significant figures. First degree algebraic equations of the type that occur frequently in general chemistry are discussed in Chapter 7. Higher degree equations, as they apply to chemical equilibrium, are considered in Chapter 8. The concept of a functional relationship (direct proportionality, inverse proportionality, linear relation) is introduced in Chapter 9. This idea is referred to again in the discussion of graphs (Chapter 10).

As the preceding paragraph implies, the application of mathematics to problems in general chemistry is stressed repeatedly throughout this book. To emphasize this connection, we have used three special devices:

— The problems at the end of each chapter are arranged in matched pairs. The first member of each pair illustrates, in simple form, a mathematical technique introduced in the chapter. The paired problem applies the same technique to a problem in general chemistry. All problems are answered in Appendix 2.

— The last chapter in the book (Chapter 11) is devoted to a general discussion of the analysis of problems in general chemistry. Emphasis is placed upon the importance of recognizing and understanding the chemical principles involved.

— A special section has been added at the end of the index to indicate where problems of a particular type (e.g., percent composition of compounds, dissociation of weak acids) are referred to.

We assume that a student using this book owns or has access to a "scientific" calculator. Chapter 1 describes how such a calculator is used in simple mathematical operations. The use of exponential notation on the calculator is covered in Chapter 4 (Exponential Numbers). Finding logarithms and antilogarithms with a calculator is discussed in Chapter 5 (Logarithms). One effect of the widespread use of calculators is to eliminate the need for extensive tables of mathematical functions in a book of this type. In particular, "log tables" have died a natural death, generally unmourned by instructors and students alike.

Many people have contributed to this book. Among these is the wife of one of the authors (WLM) whose candid criticism was usually appreciated. It is always a pleasure to work with the staff at Saunders, particularly our editor John Vondeling and his talented assistant Jeannie Shoch.

Contents

CHAPTER 1

Calculators

The hand-held electronic calculator is an essential tool nowadays for any student in general chemistry. You will use it frequently: in the laboratory, for problem assignments, and on examinations. In a department store or college book store, you are likely to find three basic types of calculators:

1. Simple arithmetical calculators, usually selling for $10 or less. These allow you to add, subtract, multiply, and divide. They may also provide for a few other simple operations. However, in general, this type of calculator is not versatile enough for your needs in general chemistry.

2. So-called "scientific" or "slide rule" calculators, typically in the $20 to $50 range. These will allow you to carry out a variety of operations in arithmetic and trigonometry. They also have some "memory" capacity, with one or more storage registers. Of particular importance in general chemistry, a scientific calculator will allow you to:

— enter and perform operations on numbers expressed in exponential notation;

— find a common (base 10) logarithm or antilogarithm (number corresponding to a given logarithm);

— raise a number to any power, n, or extract the nth root of a number.

A calculator which fits this description is entirely adequate for general chemistry.

3. Programmable calculators which are, in effect, minicomputers. These can be "programmed" for complex calculations involving a long series of steps. They are more powerful than ordinary scientific calculators and also considerably more expensive. You should buy such a calculator only if you are convinced you will need it for advanced courses. It offers no particular advantage in general chemistry.

In this chapter, we begin by looking at some of the general features of scientific calculators (Section 1.1). Then we will consider how some of the simpler arithmetical operations are carried out with a calculator (Sections 1.2–1.6). This chapter concludes with a brief discussion of a few of the things that a calculator will *not* do for you. In later chapters, we will see how a calculator can be used to deal with exponential numbers (Chapter 4) or to find logarithms and antilogarithms (Chapter 5).

1.1 GENERAL FEATURES

There are many different brands of scientific calculators on the market. These differ from one another in many trivial ways and in a few more important aspects. Among the more obvious differences are the following:

1. The manner in which successive numbers are entered into the calculator. The method used by Hewlett-Packard is, so far as we know, unique. It involves what is known as "RPN logic." Calculators made by Texas Instruments and other manufacturers, both domestic and foreign, use "algebraic notation." This involves a different series of steps, using different keys. Both systems have their advocates. It can be a bit confusing to shift suddenly from one to another: we suggest you *not* do this just before an examination.

2. The power source. Less expensive calculators generally operate on disposable batteries. Such calculators usually turn off automatically if left unused in the "on" position for some time. More expensive models have rechargeable batteries. They can be operated either from these batteries or by connecting to an electrical outlet.

One type of calculator that is becoming common nowadays uses a "liquid crystal" display. The numbers in the register appear in black on a grey background and are rather difficult to read. However, the calculator draws very little power and can be run for a year or more on a single disposable battery.

All calculators have some sort of "early warning" system that tells you when the batteries are weak. Find out what this system is and pay attention to it. The last thing you need is a "dead" calculator halfway through an examination.

3. The number of functions per key. Scientific calculators with a large number of keys may have only one function per key ([x^2] , [÷] , etc.). In more compact calculators, a single key may be used for two different functions ([x^2] *and* [÷]). One of these is marked above or below the key, the other on the key itself. If you have such a calculator, refer to the instruction manual to find out how to shift from one function to another; 5^2 and 5/2 are *not* the same thing.

4. The number of digits displayed. Calculators ordinarily carry out operations such as multiplication or division to 10 or 11 digits. Often, not all of these digits are displayed; perhaps only 8 appear. With some calculators, the number of digits in the display can be adjusted by following a procedure described in the instruction manual.

5. The manner in which entries are removed. There are always at least two different ways to "clear" or "erase" on a calculator. One is used when you press the wrong key in the middle of a calculation and want to make a single correction. Another, more powerful, clearing operation is used to erase an entire calculation and start over again. If you get desperate, most calculators can be cleared by turning them off momentarily. However, this removes everything, including any stored information.

6. The signal used to indicate "illegal" operations. When you attempt an operation that the calculator can't handle, it tells you about it. The signal may be obvious; the word [Error] may appear in the display. A more subtle error signal is a string of zeros (00000000) or nines (9.999999 99), or a blinking 0.

An error signal may mean that the operation you are trying to perform is impossible. You can't divide 5 by 0 or take the square root of −4 (try it on your calculator and see what happens). In other cases, the operation itself is a perfectly valid one. It is entirely possible to take the cube root of −8 or raise 10 to the 112th power, despite what your calculator says (try it).

1.2 ADDITION AND SUBTRACTION

Among the simplest operations to carry out on a calculator are addition and subtraction. To illustrate, suppose you want to add 23.51 and 17.83. On a calculator using algebraic notation, such as one made by Texas Instruments, you would proceed as follows:

Press	Display	
23.51	**23.51**	
[+]	**23.51**	
17.83	**17.83**	
[=]	**41.34**	(i.e., 23.51 + 17.83 = 41.34)

On a calculator with RPN logic (Hewlett-Packard), the steps would be:

Press	Display
23.51	**23.51**
[ENTER]	**23.51**
17.83	**17.83**
[+]	**41.34**

Subtraction is carried out in a similar way. Suppose you want to subtract 17.83 from 23.51:

Algebraic Notation		RPN Logic	
Press	**Display**	**Press**	**Display**
23.51	**23.51**	23.51	**23.51**
[−]	**23.51**	[ENTER]	**23.51**
17.83	**17.83**	17.83	**17.83**
[=]	**5.68**	[−]	**5.68**

With either calculator, you should find that: 23.51 − 17.83 = 5.68.

Sometimes, negative numbers are involved in addition or subtraction. You might, for example, want to add 23.51 and −17.83. This can be done on a calculator by using the "change sign" key, typically labeled as [+/−] or [CHS]. Assuming you have a calculator with algebraic notation and a +/− key, the steps would be:

Press	Display	
23.51	**23.51**	
[+]	**23.51**	
17.83	**17.83**	
[+/−]	**−17.83**	
[=]	**5.68**	(i.e., 23.51 + (− 17.83) = 5.68)

You may have noticed that the answer, 5.68, is the same as that obtained by subtracting 17.83 from 23.51. Indeed, the operation just carried out could have been simplified by taking account of the general rule that

To add a negative quantity, $-b$, subtract b

$$a + (-b) = a - b \tag{1.1}$$

An analogous rule holds for subtraction:

To subtract a negative quantity, $-b$, add b

$$a - (-b) = a + b \tag{1.2}$$

Example 1.1 Use your calculator to find the value of

a. 16.194 + 2.187 b. 16.194 − 2.187
c. 16.194 + (−2.187) d. 16.194 − (−2.187)

Solution

a. 18.381 b. 14.007
c. 16.194 − 2.187 = 14.007 (using Relation 1.1)
d. 16.194 + 2.187 = 18.381 (using Relation 1.2)

1.3 MULTIPLICATION AND DIVISION

These two operations can be carried out quite simply on a calculator. Again, the exact sequence of steps will depend upon the type of calculator you have. To illustrate, suppose you want to multiply 8.6 by 5.0:

Algebraic Notation		RPN Logic	
Press	**Display**	**Press**	**Display**
8.6	**8.6**	8.6	**8.6**
[×]	**8.6**	[ENTER]	**8.6**
5.0	**5.0**	5.0	**5.0**
[=]	**43**	[×]	**43**

We conclude that 8.6 × 5.0 = 43. In another case, suppose we want to divide 206.4 by 6.00:

Algebraic Notation		RPN Logic	
Press	**Display**	**Press**	**Display**
206.4	**206.4**	206.4	**206.4**
[÷]	**206.4**	[ENTER]	**206.4**
6.00	**6.00**	6.00	**6.00**
[=]	**34.4**	[÷]	**34.4**

In other words, 206.4/6.00 = 34.4.

Frequently you will have occasion to carry out multiplications or divisions involving negative numbers. Here, as with addition and subtraction, you can carry out the entire operation on a calculator by making use of the "change sign" key. However, it is probably simpler to apply the following general rules.

If two quantities have the same sign (both + or both −), their product or quotient is positive.

$$a \times b = (-a) \times (-b) = ab \quad (1.3)$$

$$a \div b = (-a) \div (-b) = a/b \quad (1.4)$$

If the signs of two quantities differ (+ *vs.* −), their product or quotient is negative.

$$a \times (-b) = (-a) \times b = -ab \quad (1.5)$$

$$a \div (-b) = (-a) \div b = -a/b \quad (1.6)$$

Example 1.2 Use your calculator to evaluate

a. 2.2×3.0 b. $2.2 \times (-3.0)$ c. $(-2.2) \times (-3.0)$
d. $16.4 \div 4.1$ e. $16.4 \div (-4.1)$ f. $(-16.4) \div (-4.1)$

Solution

a. 6.6 b. −6.6 (using Relation 1.5) c. 6.6 (using Relation 1.3)
d. 4.0 e. −4.0 (using Relation 1.6) f. 4.0 (using Relation 1.4)

1.4 COMBINED OPERATIONS

Most calculations in chemistry involve more than one step. Thus, you might want to add a series of numbers:

$$2.318 + 6.419 + 1.032 = ?$$

combine addition and subtraction:

$$2.318 + 6.419 - 1.032 = ?$$

divide the product of two numbers by a third number:

$$\frac{6.10 \times 2.45}{4.90} = ?$$

All of these operations and many others like them can be carried out in a continuous sequence of steps on your calculator, leading directly to the final answer. You do *not* have to start over again after each step. Consider, for example, the operation

$$\frac{6.10 \times 2.45}{4.90}$$

Here, you would proceed as follows:

Algebraic Notation		RPN Logic	
Press	**Display**	**Press**	**Display**
6.10	**6.10**	6.10	**6.10**
[×]	**6.10**	[ENTER]	**6.10**
2.45	**2.45**	2.45	**2.45**
[÷]	**14.945**	[×]	**14.945**
4.90	**4.90**	4.90	**4.90**
[=]	**3.05**	[÷]	**3.05**

We conclude that $(6.10 \times 2.45)/4.90 = 3.05$ and, incidentally, that $6.10 \times 2.45 = 14.945$. In a similar way, you should be able to show on your calculator, in one continuous operation, that

$$2.318 + 6.419 + 1.032 = 9.769$$

$$2.318 + 6.419 - 1.032 = 7.705$$

The "combined operations" required to solve certain problems may be more complex than the ones we have considered. The instruction manual that came with your calculator describes several cases of this sort and tells you how to handle them. One combined operation that arises frequently in chemistry involves a denominator with more than one term. For example, you might need to evaluate

$$\frac{0.654}{0.654 + 1.382} = ?$$

It is possible to do this in *one continuous operation*. Indeed, it can be done quite simply on a calculator using RPN logic. The sequence of steps using a calculator with algebraic notation is not obvious. However, if you read your instruction manual carefully, you should find at least two different ways to evaluate this expression. One of these involves the use of parentheses keys, labeled [(] and [)]. We won't take time to discuss this approach, but if your calculator has these keys, you should become familiar with their use. The answer you should obtain for the above ratio is 0.321 . . .*

Example 1.3 Without taking any numbers out of your calculator, evaluate

a. $\frac{(6.19 - 2.36) \times 3.26}{1.63}$ b. $\frac{2(5.31)}{1.70 + 1.84}$

Solution

a. 7.66 b. 3.00

---

*The dots which appear here and in further illustrations in this chapter indicate additional digits, perhaps as many as 7 or 8, which will appear in your calculator.

1.5 RAISING TO A POWER; EXPONENTS

Occasionally in chemistry you will need to "square" or "cube" a number. Raising to a power can be done quite simply on a calculator. Before describing this operation, it may be helpful to review briefly the symbolism involved.

Meaning of Exponents

To indicate that a quantity x is to be raised to a power, n, we write

$$x^n$$

The superscript n is referred to as an **exponent.** It tells us the number of times that x is to be multiplied by itself. Thus,

$$x^1 = x \qquad 3^1 = 3$$
$$x^2 = x \cdot x \qquad 3^2 = 3 \cdot 3 = 9$$
$$x^3 = x \cdot x \cdot x \qquad 3^3 = 3 \cdot 3 \cdot 3 = 27$$

Exponents can be negative as well as positive. We take x^{-n} to mean "*one divided by* x^n." Thus,

$$x^{-1} = 1/x^1; \qquad 3^{-1} = 1/3$$
$$x^{-2} = 1/x^2; \qquad 3^{-2} = 1/3^2 = 1/9$$
$$x^{-3} = 1/x^3; \qquad 3^{-3} = 1/3^3 = 1/27$$

The quantity $1/x$ is given a special name; it is called the **reciprocal** of x. For example, 1/3 is the reciprocal of 3, 5 is the reciprocal of 0.2, and so on.

By convention, any number raised to the zero power is taken to have the value 1. That is,

$$x^0 = 1; \qquad 3^0 = 2^0 = 6^0 = 1$$

Using the [y^x] Key

Your calculator may have a special [x^2] key which can be used to square numbers. More generally, you can use the [y^x] key to raise a number to any power. Suppose, for example, you want to carry out the operation

$$3.25^3$$

The sequence of steps is as indicated below.

Algebraic Notation		RPN Logic	
Press	**Display**	**Press**	**Display**
3.25	**3.25**	3.25	**3.25**
[y^x]	**3.25**	[ENTER]	**3.25**
3	**3**	3	**3**
[=]	**34.3 . .**	[y^x]	**34.3 . .**

Clearly, 3.25^3 is approximately 34.3

With many calculators, you cannot raise a negative number to a power. Try the following operation on your calculator:

$$(-3.25)^3$$

If you get an error signal, don't despair. You can easily perform this operation by finding 3.25^3 and applying the rules for multiplication of negative numbers (p. 5).

$$3.25^3 = (3.25)(3.25)(3.25) = 34.3 \ldots$$

$$(-3.25)^3 = (-3.25)(-3.25)(-3.25) = -34.3 \ldots$$

Example 1.4 Evaluate

a. 1.62^2 b. 1.24^5 c. $1.24^{5.05}$ d. $(-1.24)^5$

Solution

a. Using the [x^2] key if you have one, or the [y^x] key, $1.62^2 = 2.62 \ldots$
b. 2.93 . .
c. 2.96 . . Note that the exponent doesn't have to be a whole number.
d. −2.93 . . Recall that $1.24^5 = 2.93 \ldots$ Since −1.24 is multiplied by itself an odd number of times (5), the final sign must be negative.

The y^x key can also be used to raise a number to a negative power. To do this, you must use the "change sign" key. Consider, for example, the operation

$$3.25^{-3}$$

The steps involved are as follows:

Algebraic Notation		RPN Logic	
Press	**Display**	**Press**	**Display**
3.25	**3.25**	3.25	**3.25**
[y^x]	**3.25**	[ENTER]	**3.25**
3	**3**	3	**3**
[+/−]	**−3**	[CHS]	**−3**
[=]	**0.0291 . .**	[y^x]	**0.0291 . .**

Notice that, in this operation, you *must* use the "change sign" key. Be sure you know how this is labeled on your calculator ([+/−] or [CHS] or whatever). Don't confuse it with the [−] key, which is used for subtraction, *not* for changing sign from + to −.

Example 1.5 Evaluate

a. 2.33^{-2} b. $1.24^{-1.50}$

Solution

a. 0.184 . . b. 0.724 . .

Using the [1/x] Key

All scientific calculators have a [1/x] key, which can be used to take reciprocals. With this key you should be able to show that

$$1/3.000 = 0.3333\ ..$$
$$1/5.20 = 0.192\ ..$$
$$1/0.125 = 8.00$$

The [1/x] key can also be used for several other operations, where you make use of an important property of the reciprocal function. *Division by x is equivalent to multiplication by its reciprocal, 1/x.* That is,

$$\frac{a}{x} = a \cdot \frac{1}{x}$$

Thus, if you wanted to divide 206.4 by 6.00, you could proceed as follows:

Algebraic Notation		RPN Logic	
Press	**Display**	**Press**	**Display**
206.4	**206.4**	206.4	**206.4**
[×]	**206.4**	[ENTER]	**206.4**
6.00	**6.00**	6.00	**6.00**
[1/x]	**0.1666 . .**	[1/x]	**0.1666 . .**
[=]	**34.4**	[×]	**34.4**

Comparing this operation to the direct division (p. 4), you can see that it involves an extra step. Hence, it might seem the wrong way to go, unless of course your ÷ key is broken.* There are occasions, though, when it's easier to divide this way. Remember the discussion involving the calculation

$$\frac{0.654}{0.654 + 1.382} = ?$$

The easiest way to do this on a calculator with algebraic notation is to take the sum called for in the denominator, find the reciprocal of that sum using the 1/x key, and multiply by 0.654. Still another use of the 1/x key is in extracting roots (Section 1.6).

1.6 EXTRACTING A ROOT; FRACTIONAL EXPONENTS

Certain calculations in general chemistry require that you take the "square root" of a number. On rare occasions, you may need to go beyond that, taking a "cube root" or higher root.

Fractional Exponents

The square root of x is defined as the quantity which, when squared, equals x. Thus, since $3 \times 3 = 9$, it follows that 3 is the square root of 9. To indicate a square root, we may use the symbol $\sqrt[2]{\ }$ (often written simply as $\sqrt{\ }$) or the *fractional exponent* 1/2:

$$\sqrt[2]{x} = x^{1/2}$$

Other, higher roots are handled in a similar way. The cube root of x is the number which, when cubed, equals x. Thus, since $2 \times 2 \times 2 = 8$, it follows that

$$\sqrt[3]{8} = 8^{1/3} = 2$$

In general, we may refer to the nth root of x,

$$x^{1/n}$$

as the number which, when multiplied by itself n times, equals x.

---

*Unlikely but not impossible. In 1972, WLM paid $300 for a calculator which, incidentally, is less powerful than those selling today for $20. Some years ago it fell on the floor, suffering internal injuries. Now, neither the ÷ nor the 6 key work. By using the 1/x key, it is still possible to carry out "sixless" calculations. It seems a shame to discard a calculator like that.

Using the $\boxed{\sqrt{x}}$ Key

The $\boxed{\sqrt{x}}$ key on your calculator is used for square roots. Using that key, you should find that

$$\sqrt{2.000} = 1.414\ldots$$

$$\sqrt{3.000} = 1.732\ldots$$

$$\sqrt{4.000} = 2.000$$

and so on.

In using this key, you should keep one point in mind. Every positive number has two square roots. These are equal in magnitude but opposite in sign. For example, +2 and −2 are both square roots of 4:

$$2 \times 2 = 4$$

$$(-2) \times (-2) = 4$$

Using the $\boxed{y^x}$ Key

Calculators differ in the way cube roots or higher roots are extracted. Sometimes, there is a special key labeled $\boxed{\sqrt[x]{y}}$ for this purpose. In other cases, an "inverse" key is used in combination with the $\boxed{y^x}$ key. However, there is one general approach that can be used with every scientific calculator to evaluate $x^{1/n}$. It uses the reciprocal key, $\boxed{1/x}$, to supplement the $\boxed{y^x}$ key. The steps involved can be illustrated by finding the cube root of 208:

Algebraic Notation		RPN Logic	
Press	**Display**	**Press**	**Display**
208	208	208	208
$\boxed{y^x}$	208	$\boxed{\text{ENTER}}$	208
3	3	3	3
$\boxed{1/x}$	0.3333 . .	$\boxed{1/x}$	0.3333 . .
$\boxed{=}$	5.92 . .	$\boxed{y^x}$	5.92 . .

We see that $208^{1/3}$ is about 5.92.

Example 1.6 Find

a. $20.0^{1/3}$ b. $(-20.0)^{1/3}$ c. $4.21^{1/5}$

Solution

a. 2.71 . .
b. If you try this on your calculator, you will probably get an error signal. However, you should realize that, since

$$2.71 \times 2.71 \times 2.71 = 20.0$$

then, $$(-2.71) \times (-2.71) \times (-2.71) = -20.0$$

The cube root of -20.0 is -2.71.

c. 1.33 . .

1.7 LIMITATIONS OF THE CALCULATOR

An electronic calculator is a very useful tool in general chemistry, but not an all-powerful one. You should be aware of some of its limitations.

1. A CALCULATOR OFTEN GIVES AN ANSWER CONTAINING MORE DIGITS THAN ARE JUSTIFIED. To show how this comes about, suppose you are told that the distance from New York to Boston is 188 miles, to the nearest mile (i.e., 188 ± 1 mile). You want to convert this distance to kilometers, multiplying by 1.609 (1 mile = 1.609 km). Doing this on your calculator, you obtain

302.492 km

The three digits after the decimal point are really meaningless. Since the distance is known only to the nearest mile, it cannot be calculated to better than about 1 km. The distance should be reported as 302 km.

This situation arises frequently in chemistry. In Chapter 6, we will consider how to determine the uncertainty in a calculated result, using significant figures. Here we need only point out that many of the numbers that appear on your calculator will not be "significant." That is, they are not experimentally meaningful. It is for this reason that, in illustrations throughout this chapter, we have not listed all the digits that typically appear on a calculator.

2. A CALCULATOR CAN NOT BE USED TO ANALYZE A CHEMISTRY PROBLEM. You must decide how to translate the problem into a mathematical set-up that can be worked out on the calculator. Write down that set-up, perhaps as an equation, before you start pressing keys. Then the steps required are obvious, and the chances for error are greatly reduced.

3. A CALCULATOR CAN NOT SUBSTITUTE FOR A KNOWLEDGE OF THE PRINCIPLES OF MATHEMATICS. We have already seen a few examples of this. We pointed out that a calculator gives only one square root of a positive number, where there are actually two (e.g., +5 and −5 for the square roots of 25). There are many other cases where a weakness in mathematics will cost you, no matter how fancy your calculator.

PROBLEMS

These problems, and those in succeeding chapters, are arranged side-by-side in matched pairs. The problems in the left column illustrate in a simple form the mathematical techniques discussed in the chapter. The corresponding problems at the right illustrate the same techniques as they apply to general chemistry.

In this chapter, the two sets of problems differ in only one respect. Those on the left are expressed entirely in terms of numbers or letters. Those on the right are "word problems," involving quantities used in chemistry. The mathematical operations required are the same for each member of the matched pair. All the problems can be worked on your calculator.

1.1 Find the value of

a. $1.235 + 6.488$

b. $1.235 - 6.488$

c. $16 + (-18)$

d. $16 - (-18)$

1.11 The proton has a charge of +1, the electron a charge of −1. What is the charge of a

a. chlorine atom (17 protons, 17 electrons)?

b. chloride ion (17 protons, 18 electrons)?

c. sodium ion (11 protons, 10 electrons)?

1.2 Evaluate

a. $1.239 + 1.641 - 0.672$

b. $209 - (162 - 101)$

c. $-116 - 82 + 16 + 95$

1.12 For the reaction: CH_4 (g) + 2 O_2 (g) → CO_2 (g) + 2 H_2O (l), the enthalpy change, ΔH, is: $\Delta H = -393.5\text{ kJ} - 571.6\text{ kJ} - (-74.8\text{ kJ})$. Find the value of ΔH in kilojoules (kJ).

1.3 Carry out the following multiplications or divisions, using your calculator:

a. 1.69×2.38

b. $1.69 \div 2.38$

c. -1.69×2.38

d. $-1.69 \div (-2.38)$

1.13 Density is the ratio of mass to volume:

density = mass/volume

a. What is the density of a sample weighing 1.648 g with a volume of 1.235 cm^3?

b. A sample of aluminum has a volume of 12.6 cm^3. The density of aluminum is 2.70 g/cm^3. What is the mass, in grams, of the sample?

c. What is the volume of a piece of aluminum weighing 12.7 g? (Take the density to be 2.70 g/cm^3 and note that volume = mass/density.)

1.4 Find the value of

a. $1.665 \times 2.104 \times (-1.892)$

b. $12.069 \div 11.152$

c. $\dfrac{6.124/3.823}{9.187}$

1.14 The concentration of a solution may be expressed as molarity (M):

$$M = \frac{\text{no. moles solute}}{\text{no. liters solution}} = \frac{\text{no. grams solute/GMM}}{\text{no. liters solution}}$$

where GMM stands for the molar mass in grams. What is the molarity of a solution containing

a. 1.60 mol of methyl alcohol in 3.44 ℓ of solution?

b. 1.60 g of methyl alcohol (GMM = 32.0 g/mol) in 3.44 ℓ of solution?

1.5 Evaluate

a. $1.392 - 2(0.114)$

b. $(2.05)(-4.40) + 1.61$

c. $-9.61 + 250(0.0034)$

1.6 Determine the value of

a. $\frac{1.619 \times 2.468}{3.291}$

b. $\frac{-1.84 \times 2.97 \times 6.12}{2.42}$

c. $\frac{1.02 \times 0.0821 \times 297}{1.19}$

1.7 Find the value of

a. $\frac{1.69 + 2.38}{6.98}$

b. $\frac{1.20}{3.02(1.88)}$

c. $\frac{0.451}{0.451 + 1.324}$

1.8 Carry out the following operations:

a. $(1.62)^2$ *b*. $(4.18)^{-1}$

c. $(3.18)^{-3}$ *d*. $4\pi(1.92)^3$

1.9 Find the value of

a. $(4.42)^{1/2}$

b. $(8.18)^{1/3}$

c. $(-8.18)^{1/3}$

1.15 The free energy change, ΔG, in a reaction can be calculated from the equation

$$\Delta G = \Delta H - T\Delta S$$

where ΔH is the enthalpy change, T is the absolute temperature, and ΔS is the entropy change. For a certain reaction, $\Delta H = -8.10$ kJ, $T = 300$ K, $\Delta S = 0.0034$ kJ/K. Calculate ΔG in kilojoules (kJ).

1.16 The volume, V, of an ideal gas is given by

$$V = \frac{nRT}{P}$$

where n is the number of moles, T is the temperature, P is the pressure, and R is a constant, 0.0821 $\ell \cdot$ atm/(mol $\cdot$ K). Calculate V in liters (ℓ) when $n = 1.02$ mol, $T = 298$ K, $P = 1.18$ atm.

1.17 The mole fraction, X_2, of a solute in a solution is given by the equation

$$X_2 = \frac{n_2}{n_2 + n_1}$$

where n_2 is the number of moles of solute and n_1 is the number of moles of solvent. Calculate X_2 when

a. $n_2 = n_1 = 1.00$

b. $n_2 = 0.451$; $n_1 = 1.324$

1.18 The volume, V, of a (spherical) atom is given by the expression

$$V = 4\pi r^3/3$$

where r is the atomic radius. Taking the atomic radius of cesium to be 0.262 nm, calculate the volume of a cesium atom in cubic nanometers (nm^3).

1.19 The average velocity, u, of a gas molecule is given by

$$u = \left(\frac{3RT}{\text{KMM}}\right)^{1/2}$$

where T is the temperature, KMM is the molar mass in kilograms, and R is a constant. Taking $R = 8.31$, $T = 300$, KMM $= 0.0320$, calculate u (the units of u are meters per second).

1.10 Two quantities x and y are related by the equation

$$x = 2y^3$$

a. Find x when $y = 6.07$

b. Find y when $x = 2.40$

1.20 The equilibrium constant for the reaction

$$2\ H_2O\ (l) \rightleftharpoons 2\ H_2\ (g) + O_2\ (g)$$

is given by the expression

$$K = [H_2]^2\ x[O_2]$$

where $[H_2]$ and $[O_2]$ are the equilibrium concentrations of hydrogen and oxygen. Calculate

a. K if $[H_2] = [O_2] = 0.10$

b. $[H_2]$ if $K = 4.0$ and $[O_2] = 1.5$

CHAPTER 2

Unit Conversions

In a sense, all problems in general chemistry involve a "conversion" from a given quantity to another quantity for which you are asked to solve. Sometimes, little more than a change in units is involved, perhaps from kilograms to grams or grams to moles. Such problems are readily solved by the "conversion factor" approach, which is the subject of this chapter. Indeed, as we will see, this approach can be extended to solve more complex problems, where more than one conversion is required.

To many students, the conversion factor approach is new and, hence, automatically suspect. Applied to simple problems, it may seem awkward or artificial at first. As you gain more experience with the approach, you will discover two of its advantages. In the first place, it provides a convenient and general method of setting up solutions for a wide variety of problems in general chemistry. Equally important, it encourages you to analyze a problem to decide upon the path that you will follow to solve it.

2.1 SIMPLE CONVERSIONS

To illustrate the conversion factor approach, let us start with a simple problem. Suppose we are asked to convert a distance of 113 miles to kilometers. To do this, we need a relation between these two units. That relation is

$$1.609 \text{ km} = 1 \text{ mile} \tag{2.1}$$

In other words, a distance of one mile is equal in magnitude to 1.609 km. Suppose that now we divide both sides of this equation by 1 mile:

$$\frac{1.609 \text{ km}}{1 \text{ mile}} = \frac{1 \text{ mile}}{1 \text{ mile}} = 1$$

The quotient 1.609 km/1 mile is called a **conversion factor.** As you can see, this quotient is numerically equal to 1. Hence, if we multiply 113 miles by 1.609 km/1 mile, we do not change the value of the distance. We do, however, change its units, converting from miles to kilometers:

$$113 \cancel{\text{ miles}} \times \frac{1.609 \text{ km}}{1 \cancel{\text{ mile}}} = 182 \text{ km*}$$

*In this and all other calculations, we follow the rules of significant figures (Chapter 6).

The relation given by Equation 2.1 can be used equally well to convert a distance in kilometers, let us say 243 km, to miles. In this case, we divide both sides of the equation by 1.609 km to obtain another, related, conversion factor:

$$\frac{1 \text{ mile}}{1.609 \text{ km}} = 1$$

Multiplying 243 km by the *conversion factor* 1 mile/1.609 km converts the distance from kilometers to miles:

$$243 \cancel{\text{km}} \times \frac{1 \text{ mile}}{1.609 \cancel{\text{km}}} = 151 \text{ miles}$$

Notice that a single relation (e.g., 1.609 km = 1 mile) gives us two conversion factors, both of which are equal to unity. One of these, 1.609 km/1 mile, allows us to convert from miles to kilometers. The other, 1 mile/1.609 km, is used when we want to go in the opposite direction, from kilometers to miles. In general, to make a conversion, we choose the factor that allows us to cancel the unit we want to get rid of:

$$\text{initial quantity} \times \text{conversion factor} = \text{desired quantity}$$

The conversion factor approach is particularly useful in dealing with units that may be unfamiliar to you (Example 2.1).

Example 2.1 In the United States, atmospheric pressure is most often reported in "inches of mercury" (inch Hg). In Canada and other countries using the International System of Units (SI), the preferred unit is the kilopascal (kPa). The relationship between these two pressure units is

$$1 \text{ kPa} = 0.2954 \text{ inch Hg}$$

Convert a pressure of 29.65 inch Hg to kilopascals.

Solution

To make the conversion, we need a quotient in which kilopascals appear in the numerator and inches of mercury in the denominator. This quotient is

$$\frac{1 \text{ kPa}}{0.2954 \text{ inch Hg}}$$

Multiplying 29.65 inch Hg by this quotient,

$$29.65 \text{ inch Hg} \times \frac{1 \text{ kPa}}{0.2954 \text{ inch Hg}} = 100.4 \text{ kPa}$$

Here, as in all problems of this type, you should include the units (e.g., inches of mercury, kilopascals) in setting up the problem. If the correct conversion factor (1 kPa/0.2954 inch Hg) is used, the original unit (inch Hg) will cancel to give the answer in the desired unit (kPa). Suppose, however, that you get the conversion factor upside down

(0.2954 inch Hg/1 kPa). Then your answer will come out in nonsensical units, (inch Hg)^2/kPa, indicating a mistake in reasoning.

Sometimes the two quantities which appear in a conversion factor are *equivalent* to one another rather than equal. Suppose, for example, that platinum is selling for \$12 a gram. Strictly speaking, we wouldn't want to say that one gram of platinum "equals" \$12; the two quantities represent quite different things, like apples and oranges. However, it would be legitimate to write the relation

$$\$12 \simeq 1 \text{ g Pt} \qquad (2.2)$$

Here, the symbol $\simeq$ means **"is equivalent to."** For all practical purposes, the equivalence sign can be treated as an equals sign (Example 2.2).

Example 2.2 Using Equation 2.2, determine

a. the cost of 2.5 g of platinum
b. the mass of platinum that can be bought for \$42

Solution

a. We need a conversion factor to go from grams of platinum to dollars. To obtain it, we divide both sides of Equation 2.2 by one gram of platinum:

$$\frac{\$12}{1 \text{ g Pt}} = \frac{1 \text{ g Pt}}{1 \text{ g Pt}} = 1$$

(Note that the equivalence sign is treated as if it were an equals sign). Now we use this factor to find the cost of 2.5 g of platinum:

$$2.5 \text{ g Pt} \times \frac{\$12}{1 \text{ g Pt}} = \$30$$

b. Here, we want to go in the opposite direction, from dollars to grams of platinum. The conversion factor required is: 1 g Pt/\$12. Setting up the calculation,

$$\$42 \times \frac{1 \text{ g Pt}}{\$12} = 3.5 \text{ g Pt}$$

We could buy 3.5 g of Pt for \$42.

As we will see later in this chapter, "equivalences" are very common in chemistry. You will come across the $\simeq$ sign frequently throughout the remainder of this chapter. Remember: *Mathematically, it can be treated as an equals sign.*

2.2 MULTIPLE CONVERSIONS

Sometimes the conversion factor that is required to solve a problem is not available directly. Instead, it must be calculated from other, simpler factors. As an example, suppose you were required to convert a volume of 2.12 inch^3 to cubic centimeters, given only the relationship

$$1 \text{ inch} = 2.54 \text{ cm} \tag{2.3}$$

Clearly, to carry out this conversion, you need a relation between *cubic* inches and *cubic* centimeters. This relation is readily obtained by cubing both sides of Equation 2.3:

$$(1 \text{inch})^3 = (2.54 \text{ cm})^3$$

$$1 \text{ inch}^3 = (2.54)^3 \text{ cm}^3 = 16.4 \text{ cm}^3 \tag{2.4}$$

Using the conversion factor obtained from Equation 2.4, we find that 2.12 inch^3 is equal to 34.8 cm^3.

$$2.12 \text{ inch}^3 \times \frac{16.4 \text{ cm}^3}{1 \text{ inch}^3} = 34.8 \text{ cm}^3$$

Frequently, in working a problem, we have to make two or more successive conversions. Suppose, for example, we are asked to convert an energy change expressed in kilocalories, 1.64 kcal, to joules, the preferred SI energy unit. The relations that we have to work with are

$$1 \text{ kcal} = 10^3 \text{ cal}; \qquad 1 \text{ cal} = 4.184 \text{ J}$$

To make this conversion, it would seem reasonable to first convert 1.64 kcal to calories:

$$1.64 \text{ kcal} \times \frac{10^3 \text{ cal}}{1 \text{ kcal}} = 1.64 \times 10^3 \text{ cal}$$

and then to joules:

$$1.64 \times 10^3 \text{ cal} \times \frac{4.184 \text{ J}}{1 \text{ cal}} = 6.86 \times 10^3 \text{ J}$$

Ordinarily, we would set up this conversion on a single line, without solving for an intermediate answer. That is, we would write

$$1.64 \text{ kcal} \times \frac{10^3 \text{ cal}}{1 \text{ kcal}} \times \frac{4.184 \text{ J}}{1 \text{ cal}} = 6.86 \times 10^3 \text{ J}$$

Example 2.3 illustrates a slightly different type of two-step conversion.

Example 2.3 The density of mercury, in grams per cubic centimeter, is 13.6 g/cm^3. Express its density in kilograms per cubic meter, given:

$$1 \text{ kg} = 10^3 \text{ g}; \qquad 1 \text{ m}^3 = 10^6 \text{ cm}^3$$

Solution

Let's proceed one step at a time, first converting grams to kilograms and then cubic centimeters to cubic meters. To keep track of what we're doing, we will list the steps:

(1) grams to kilograms (1 kg = 10^3 g)
(2) cubic centimeters to cubic meters (1 m^3 = 10^6 cm^3)

$$\text{density Hg} = 13.6\,\frac{\text{g}}{\text{cm}^3} \times \underset{(1)}{\frac{1 \text{ kg}}{10^3 \text{ g}}} \times \underset{(2)}{\frac{10^6 \text{ cm}^3}{1 \text{ m}^3}} = 13.6 \times 10^3 \text{ kg/m}^3$$

$$= 1.36 \times 10^4 \text{ kg/m}^3$$

Occasionally, a problem may require three or even four conversions. Such problems can be handled exactly like the one just worked, setting up the various steps in succession. To avoid getting lost en route, it's a good idea to map out each step before you start.

Example 2.4 Suppose you drive your car a distance of 150 km. How many liters of gasoline will be required, if the car gets 22.6 miles per gallon? (1 mile = 1.609 km; 1 gallon = 3.78 ℓ.)

Solution

Reading the problem carefully, we see that we are asked, in effect, to convert a distance, 150 km, into a volume of gasoline, expressed in liters. We might logically do this in three steps:

(1) Convert 150 km to miles (1 mile = 1.609 km)
(2) Convert miles to gallons of gasoline, knowing that 22.6 miles $\triangleq$ 1 gallon where the symbol $\triangleq$ means "is equivalent to."
(3) Convert gallons to liters (1 gallon = 3.78 ℓ)

The set-up is:

$$150 \text{ km} \times \underset{(1)}{\frac{1 \text{ mile}}{1.609 \text{ km}}} \times \underset{(2)}{\frac{1 \text{ gallon}}{22.6 \text{ mile}}} \times \underset{(3)}{\frac{3.78\ \ell}{1 \text{ gallon}}} = 15.6\ \ell$$

Notice that so far as the set-up of the problem is concerned, the equivalence (2) is handled exactly like the equalities (1) and (3).

2.3 APPLICATIONS IN GENERAL CHEMISTRY

The conversion factor approach is useful in solving a variety of problems in general chemistry. Many of these involve a basic counting unit used over and over again in chemistry: the **mole.** A mole, which is always associated with a chemical symbol or formula, represents Avogadro's number (6.022×10^{23}) of items. The kind of "item" is indicated by the nature of the symbol or formula. Thus, we have

$$1 \text{ mol Ag} = 6.022 \times 10^{23} \text{ Ag atoms}$$
$$1 \text{ mol } H_2 = 6.022 \times 10^{23}\ H_2 \text{ molecules}$$
$$1 \text{ mol } H_2O = 6.022 \times 10^{23}\ H_2O \text{ molecules}$$

The mass of a mole, in grams, is found by adding up the atomic masses of the atoms present:

$$1 \text{ mol Ag} = 107.87 \text{ g Ag}$$
$$1 \text{ mol } H_2 = 2(1.008 \text{ g}) = 2.016 \text{ g}H_2$$
$$1 \text{ mol } H_2O = 2(1.008 \text{ g}) + 16.000 \text{ g} = 18.016 \text{ g } H_2O$$

Conversions between moles and numbers of particles (atoms, molecules) or masses in grams are readily made. All that is required is that you know what is meant by a mole (Example 2.5).

Example 2.5 Determine the

a. number of molecules in 1.24 mol of H_2O
b. mass in grams of 0.198 mol of H_2O
c. number of moles of H_2O in 16.4 g

Solution

a. The required relation is

$$1 \text{ mol } H_2O = 6.022 \times 10^{23} \text{ molecules } H_2O$$

This gives us the conversion factor we need to go from moles to molecules:

$$1.24 \text{ mol } H_2O \times \frac{6.022 \times 10^{23} \text{ molecules } H_2O}{1 \text{ mol } H_2O} = 7.47 \times 10^{23} \text{ molecules } H_2O$$

b. We use the relation

$$1 \text{ mol } H_2O = 18.016 \text{ g } H_2O$$

Since we want to go from moles to grams, we have

$$0.198 \text{ mol } H_2O \times \frac{18.016 \text{ g } H_2O}{1 \text{ mol } H_2O} = 3.57 \text{ g } H_2O$$

c. Here again, we use the relation in (b). However, since we want to go from grams to moles, we use the conversion factor: 1 mol/18.016 g.

$$16.4 \text{ g } H_2O \times \frac{1 \text{ mol } H_2O}{18.016 \text{ g } H_2O} = 0.910 \text{ mol } H_2O$$

Notice that the relation between moles and molecules (1 mol = 6.022×10^{23} molecules) is the same for all substances. The answer to Example 2.5, part a, would have been the same for H_2, CO_2, or any other molecular substance. In contrast, the relation between moles and grams differs from one substance to another. One mole of H_2O weighs 18.016 g, whereas one mole of H_2 weighs only 2.016 g. The answers to Example 2.5, parts b and c, would have been quite different if we had been dealing with H_2 instead of H_2O.

Perhaps the most common use of the conversion factor approach in chemistry occurs in connection with balanced equations. Here we use conversion factors to relate the amounts of two different substances involved in a reaction. These amounts may be expressed either in moles or grams. Mole-mole, mole-gram, or gram-gram conversions involving balanced equations require that you

—be thoroughly familiar with the conversion factor approach;

—know what a mole is and how to obtain its mass in grams;

—realize that the coefficients in a balanced equation represent relative numbers of moles. As an example, the balanced equation

$$2\ H_2\ (g) + O_2\ (g) \rightarrow 2\ H_2O\ (l)$$

tells us that 2 mol H_2 reacts with 1 mol O_2 to form 2 mol H_2O. Putting it another way, in this reaction,

$$2\ \text{mol}\ H_2 \simeq 1\ \text{mol}\ O_2 \simeq 2\ \text{mol}\ H_2O$$

where the $\simeq$ means "is chemically equivalent to." Here, two moles of H_2 are equivalent to one mole of O_2 in the sense that these two quantities react with each other. Both these quantities are chemically equivalent to two moles of H_2O, since that amount of water is produced in the reaction.

Example 2.6 The combustion of butane gas, C_4H_{10}, is represented by the balanced equation

$$2\ C_4H_{10}\ (g) + 13\ O_2\ (g) \rightarrow 8\ CO_2\ (g) + 10\ H_2O\ (l)$$

Using this equation, calculate

a. the number of moles of CO_2 formed by the combustion of 1.42 mol of C_4H_{10}
b. the number of grams of H_2O formed from 1.42 mol of C_4H_{10}
c. the mass in grams of O_2 required to form 21.0 g of CO_2.

Solution

a. The conversion factor required here follows directly from the coefficients of the balanced equation. In this reaction, 2 mol of butane produce 8 mol of carbon dioxide, that is,

$$2\ \text{mol}\ C_4H_{10} \simeq 8\ \text{mol}\ CO_2$$

Starting with 1.42 mol of C_4H_{10}, we must then obtain

$$1.42\ \text{mol}\ C_4H_{10} \times \frac{8\ \text{mol}\ CO_2}{2\ \text{mol}\ C_4H_{10}} = 5.68\ \text{mol}\ CO_2$$

b. Here we need to convert moles of C_4H_{10} to grams of H_2O. A logical two-step path would be

(1) moles C_4H_{10} to moles H_2O $\quad$ ($2 \text{ mol } C_4H_{10} \simeq 10 \text{ mol } H_2O$)

(2) moles H_2O to grams H_2O $\quad$ ($1 \text{ mol } H_2O = 18.0 \text{ g } H_2O$)

The set-up is

$$\text{mass } H_2O = 1.42 \text{ mol } C_4H_{10} \times \underset{(1)}{\frac{10 \text{ mol } H_2O}{2 \text{ mol } C_4H_{10}}} \times \underset{(2)}{\frac{18.0 \text{ g } H_2O}{1 \text{ mol } H_2O}} = 128 \text{ g } H_2O$$

c. Extending the analysis of parts a and b, it would appear that a three-step path is in order:

(1) grams CO_2 to moles CO_2 $\quad$ ($1 \text{ mol } CO_2 = 44.0 \text{ g } CO_2$)

(2) moles CO_2 to moles O_2 $\quad$ ($13 \text{ mol } O_2 \simeq 8 \text{ mol } CO_2$)

(3) moles O_2 to grams O_2 $\quad$ ($1 \text{ mol } O_2 = 32.0 \text{ g } O_2$)

Starting with 21.0 g of CO_2, the conversion is

$$\text{mass } O_2 = 21.0 \text{ g } CO_2 \times \underset{(1)}{\frac{1 \text{ mol } CO_2}{44.0 \text{ g } CO_2}} \times \underset{(2)}{\frac{13 \text{ mol } O_2}{8 \text{ mol } CO_2}} \times \underset{(3)}{\frac{32.0 \text{ g } O_2}{1 \text{ mol } O_2}} = 24.8 \text{ g } O_2$$

A slightly different approach is available for multistep conversions of the type shown in Example 2.6, parts b and c. Here, we calculate in advance a single factor which will accomplish the required conversion in a single step. In part b, we could have reasoned that since $2 \text{ mol } C_4H_{10} \simeq 10 \text{ mol } H_2O$, and one mole of water weighs 18.0 g,

$$2 \text{ mol } C_4H_{10} \simeq 10(18.0 \text{ g } H_2O) = 180 \text{ g } H_2O$$

Now we have the conversion factor we need to go directly from 1.42 mol of C_4H_{10} to grams of H_2O. The single-step conversion is

$$1.42 \text{ mol } C_4H_{10} \times \frac{180 \text{ g } H_2O}{2 \text{ mol } C_4H_{10}} = 128 \text{ g } H_2O$$

We can apply the same sort of reasoning in part c. Here again we start with the relationship given by the coefficients of the balanced equation

$$13 \text{ mol } O_2 \simeq 8 \text{ mol } CO_2$$

We now use the fact that one mole of O_2 weighs 32.0 g whereas one mole of CO_2 weighs 44.0 g. This leads directly to the relationship we need, that between grams of O_2 and grams of CO_2:

$$13(32.0 \text{ g } O_2) \simeq 8(44.0 \text{ g } CO_2)$$

$$416 \text{ g } O_2 \simeq 352 \text{ g } CO_2$$

Now, with a single conversion, we find the number of grams of O_2 required to form 21.0 g of CO_2:

$$21.0 \text{ g } CO_2 \times \frac{416 \text{ g } O_2}{352 \text{ g } CO_2} = 24.8 \text{ g } O_2$$

The approach just described leads, of course, to the same answer as that given in Example 2.6. Use whichever approach appeals to you, that is, whichever is easier for you to follow and apply.

The conversion factor approach is readily extended to *thermochemical equations,* where the heat flow, ΔH, is specified, usually in kilojoules (kJ). Consider, for example,

$$2 \text{ } H_2 \text{ (g)} + O_2 \text{ (g)} \rightarrow 2 \text{ } H_2O \text{ (l)}; \qquad \Delta H = -571.6 \text{ kJ}$$

We interpret this thermochemical equation to mean that when 2 mol H_2 reacts with 1 mol O_2 to form 2 mol H_2O, the enthalpy change, ΔH, is −571.6 kJ. Putting it another way, 571.6 kJ of heat is *evolved* in this reaction. In terms of "equivalences,"

$$2 \text{ mol } H_2 \simeq 1 \text{ mol } O_2 \simeq 2 \text{ mol } H_2O \simeq -571.6 \text{ kJ}$$

These relations lead to conversion factors which allow us to relate ΔH for a reaction to amounts of products or reactants (Example 2.7).

Example 2.7 Using the relations just given, calculate

a. ΔH when 1.00 mol of H_2O forms

b. the mass in grams of H_2 required to produce 1.00 kJ of heat (i.e., to make $\Delta H = -1.00$ kJ)

Solution

a. We need a relation between moles of H_2O and heat flow (ΔH). That relation is

$$2 \text{ mol } H_2O \simeq -571.6 \text{ kJ}$$

Using this relation to obtain the appropriate conversion factor, we have

$$\Delta H = 1.00 \text{ mol } H_2O \times \frac{-571.6 \text{ kJ}}{2 \text{ mol } H_2O} = -285.8 \text{ kJ}$$

b. Here we want to go from heat flow (−1.00 kJ) to grams of H_2. A logical two-step path would be

(1) heat flow to moles H_2 $\quad$ ($2 \text{ mol } H_2 \simeq -571.6 \text{ kJ}$)

(2) moles H_2 to grams H_2 $\quad$ ($1 \text{ mol } H_2 = 2.016 \text{ g } H_2$)

Hence,

$$\text{mass } H_2 = -1.00 \text{ kJ} \times \underset{(1)}{\frac{2 \text{ mol } H_2}{-571.6 \text{ kJ}}} \times \underset{(2)}{\frac{2.016 \text{ g } H_2}{1 \text{ mol } H_2}} = 0.00705 \text{ g } H_2$$

We conclude that only about 7×10^{-3} g of hydrogen needs to be burned to produce a kilojoule of heat.

Sometimes in chemistry, a conversion factor is "hidden" in the statement of a problem. Finding the conversion factor may suggest a simple way to work the problem. This is the case with the solubility calculations called for in Example 2.8.

Example 2.8 At 80° C, the solubility of potassium nitrate, KNO_3, is 169 g per 100 g of water. At 20° C, the solubility is only 32 g/100 g of water. Using a conversion factor approach, calculate

a. the amount of water required to dissolve 120 g of KNO_3 at 80° C.

b. the amount of KNO_3 that will dissolve in 62 g of water at 20° C.

Solution

The saturated solution at 80° C contains 169 g of KNO_3 in 100 g of water. In that sense, these two quantities are equivalent to each other:

$$\text{at } 80^\circ \text{ C:} \qquad 169 \text{ g } KNO_3 \triangleq 100 \text{ g water}$$

Similarly,

$$\text{at } 20^\circ \text{ C:} \qquad 32 \text{ g } KNO_3 \triangleq 100 \text{ g water}$$

These relations lead directly to the conversion factors we need to solve the problem.

a. We want to convert 120 g of KNO_3 to the equivalent amount of water at 80° C:

$$120 \text{ g } KNO_3 \times \frac{100 \text{ g water}}{169 \text{ g } KNO_3} = 71.0 \text{ g water}$$

Hence, 71.0 g of water is required to dissolve 120 g of KNO_3 at 80° C.

b. Here we are going in the opposite direction, so the appropriate conversion factor is 32 g KNO_3/100 g water:

$$62 \text{ g water} \times \frac{32 \text{ g } KNO_3}{100 \text{ g water}} = 20 \text{ g } KNO_3$$

We conclude that 20 g of KNO_3 will dissolve in 62 g of water at 20° C.

We could cite many other examples where the conversion factor approach is useful in general chemistry. However, it may be best to let you find these for yourself. Instead, it may be appropriate to end this chapter by expressing a couple of reservations about the use of conversion factors.

1. While the conversion factor approach is a valuable method of analyzing problems in general chemistry, it is by no means the only approach. Some students become so convinced of its merits that they try to apply this approach to every problem they meet. Don't try to force a fit in this way. If, after reflection, the problem does not seem to lend itself to this approach, try another. Many problems can be analyzed most simply in terms of an algebraic equation (Chapter 7) relating the quantity you are given to that for which you are asked to solve.

2. Remember that the conversion factor approach is really a way of analyzing a problem. It is not a "magic formula" for solving problems. If you try to use it

mechanically, it can become a rote method, no better than any other. In chemistry (and in just about everything else that we can think of), there is no substitute for understanding what you are doing.

PROBLEMS

2.1 As you know, the word "dozen" is used to represent 12 objects. Find

a. the number of eggs in 16 dozen

b. the number of dozens in 594 apples

2.2 A crucible weighs 40.6 g. Using the relations

$$1 \text{ kg} = 10^3 \text{ g} = 2.205 \text{ lb}$$

express the mass of the crucible in

a. kilograms *b.* pounds

2.3 The prices of ground beef, roast pork, and sirloin steak in a supermarket are \$1.79, \$2.12, and \$2.87 a pound in that order. What is the cost of a 120 g portion of

a. ground beef *b.* roast pork

c. sirloin steak

(1 lb = 453.6 g)

2.4 To make antifreeze for use in very cold climates, 7.0 ℓ of ethylene glycol is used for 3.0 ℓ of water.

a. How much water should be added to 5.0 ℓ of ethylene glycol to make antifreeze?

b. How much ethylene glycol is required for 4.0 ℓ of water?

2.5 The barometric pressure on a certain day is reported to be 29.71 inch Hg. Express this in

a. atmospheres *b.* kilopascals

c. millimeters of mercury

(1 atm = 760 mm Hg = 101.3 kPa = 29.92 inch Hg)

2.6 The density of O_2 at 25° C and 1.00 atm is 1.31 g/ℓ. Express this in

a. grams per cubic centimeter

b. kilograms per cubic meter

(1 kg = 10^3 g; 1 m^3 = 10^3 ℓ = 10^6 cm^3)

2.11 Knowing that 1 mol = 6.022×10^{23} molecules, find

a. the number of molecules in 2.14 mol

b. the number of moles in 3.19×10^{22} molecules

2.12 For carbon dioxide,

$$1 \text{ mol} = 6.022 \times 10^{23} \text{ molecules} = 44.01 \text{ g}$$

Convert 4.06 mol of CO_2 to

a. grams *b.* molecules

2.13 Consider the three hydrocarbons CH_4, C_2H_2, and C_6H_6. Taking the atomic masses of C and H to be 12.01 and 1.008, calculate

a. the mass in grams of one mole of CH_4; C_2H_2; C_6H_6

b. the number of moles in 122 g of each of these substances

2.14 The solubility of silver nitrate, $AgNO_3$, is 952 g/100 g of water at 100° C and 222 g/100 g of water at 20° C.

a. How much water is required to dissolve 206 g of $AgNO_3$ at 100° C?

b. If the solution in (*a*) is cooled to 20° C (the amount of water stays constant), how much silver nitrate stays in solution? (The excess $AgNO_3$ crystallizes out of solution.)

2.15 The gas law constant, *R*, is 0.0821 ℓ · atm/(mol · K). Express *R* in

a. $\dfrac{\ell \cdot \text{kPa}}{\text{mol} \cdot \text{K}}$ *b.* $\dfrac{\ell \cdot \text{mm Hg}}{\text{mol} \cdot \text{K}}$ *c.* $\dfrac{\text{cm}^3 \cdot \text{atm}}{\text{mol} \cdot \text{K}}$

(Use the relations in Problem 2.5 and the fact that 1000 cm^3 = 1 ℓ.)

2.16 The concentration of sulfuric acid in "acid rain" can be as high as 1.1×10^{-1} g/ℓ. Express this concentration in

a. grams per cubic centimeter

b. kilograms per cubic meter

c. moles per liter (1 mol H_2SO_4 = 98.1 g)

2.7 A good time for the 100-yd dash is ten seconds. Express this speed (3 significant figures) in kilometers per hour. (1 mile = 1760 yd = 1.609 km; 1 h = 60 min; 1 min = 60 s)

2.8 To make a cream sauce, a recipe calls for 2 tablespoons butter, 3/2 tablespoons flour, and 1 cup milk: 2 tbsp butter + 3/2 tbsp flour + 1 cup milk → cream sauce:

a. How many tablespoons of butter are required for 7.0 cups of milk?

b. How many cups of milk are required for 2.0 tablespoons of flour?

c. How many cups of butter are needed for 6.0 tablespoons of flour (16 tbsp = 1 cup)?

2.9 A recipe for making daiquiris calls for adding one package (0.58 oz) of powdered mix to 1.50 oz of rum and 1.50 oz of water:

0.58 oz mix + 1.50 oz rum + 1.50 oz water → 1 daiquiri

a. What is the mass in ounces of one daiquiri?

b. What mass of rum is required to form six daiquiris?

c. How many ounces of mix are required for 6.6 ounces of rum?

d. How many daiquiris can be made from 12 oz of rum?

2.10 To heat 1.00 kg of water from 25° C to 100° C in an open saucepan, 314 kJ of heat must be absorbed.

a. How much heat must be supplied to heat 742 g of water from 25 to 100° C?

b. How much water can be heated from 25 to 100° C by the absorption of 1.00 kJ?

2.17 According to the kinetic theory of gases, the average velocity of an O_2 molecule at 25° C is 482 m/s. Express this in

a. kilometers per hour

b. centimeters per day

2.18 Given the balanced equation

$$4\ NH_3\ (g) + 5\ O_2\ (g) \rightarrow 4\ NO\ (g) + 6\ H_2O\ (l)$$

and taking the atomic masses of N, H, and O to be 14.0, 1.0, and 16.0, calculate

a. the number of moles of O_2 required to react with 1.51 mol of NH_3

b. the number of grams of H_2O produced from 0.282 mol of NH_3

c. the number of moles of NH_3 required to form 6.40 g of NO

d. the mass in grams of NO formed from 9.80 g of O_2

2.19 For the reaction

$$2\ Bi^{3+}(aq) + 3\ S^{2-}(aq) \rightarrow Bi_2S_3(s)$$
(AM Bi = 209.0, S = 32.1)

determine

a. the number of moles of Bi_2S_3 formed from 1.61 mol of Bi^{3+}

b. the number of grams of Bi_2S_3 formed from 2.11 mol of S^{2-}

c. the number of grams of Bi^{3+} required to react with 1.00 g of S^{2-}

2.20 Given the thermochemical equation

$$CH_4\ (g) + 2\ O_2\ (g) \rightarrow CO_2\ (g) + 2\ H_2O\ (l);\ \Delta H = -890.3\ kJ$$

Calculate

a. ΔH when 0.682 mol of CH_4 burns

b. the number of grams of CH_4 (MM = 16.04) that must be burned to evolve 10.0 kJ of heat (i.e., $\Delta H = -10.0$ kJ)

CHAPTER 3

Percent

The concept of percent is a familiar one. A TV announcer tells us that a candidate received 51% of the vote in an election. We read in the morning paper that the population of the earth is increasing 2% each year. Most of us feel that we know what these numbers mean and how to use them in calculations. Just to make sure, the first section of this chapter reviews the arithmetic of percentages.

Many of the principles of chemistry are expressed in the language of percent. For example, the atomic mass of an element can be calculated from the mole percents of the isotopes present (Example 3.7). The simplest formula of a compound can be obtained from the mass percents of the elements present (Example 3.10). These and other applications of the percent concept in general chemistry are discussed in Section 3.2.

3.1 BASIC RELATIONS

Percent means literally "parts per hundred." A candidate who receives 51% of the vote in an election gets 51/100 of the total vote. When we say that air contains 78 mole percent of N_2, we mean that, in 100 mol of air, there is 78 mol of N_2.

The percent of a component A in a sample can be determined by comparing the amount of A with the total amount of sample. The general relation is:

$$\% \text{ of } A = \frac{\text{amount of } A}{\text{total amount}} \times 100 \qquad (3.1)$$

The percents of all the components ($A, B, C, \ldots$) in a sample must add up to 100:

$$\% \text{ of } A + \% \text{ of } B + \% \text{ of } C + \cdots = 100 \qquad (3.2)$$

This equation tells us, for example, that if we add up the percents of the votes obtained by all the candidates in an election, the sum must be 100.

Example 3.1 Analysis of an iron-sulfur mixture weighing 12.50 g shows that it contains 2.70 g of iron. What are the mass percents of iron and sulfur?

Solution

To find the percent of iron, we apply Equation 3.1 directly:

$$\% \text{ of iron} = \frac{\text{mass iron}}{\text{total mass sample}} \times 100 = \frac{2.70 \text{ g}}{12.50 \text{ g}} \times 100 = 21.6$$

To obtain the percent of sulfur, we could first obtain its mass:

$$\text{mass sulfur} = \text{total mass sample} - \text{mass iron}$$

$$= 12.50 \text{ g} - 2.70 \text{ g} = 9.80 \text{ g}$$

Then we could proceed as with iron:

$$\% \text{ of sulfur} = \frac{9.80 \text{ g}}{12.50 \text{ g}} \times 100 = 78.4$$

A simpler approach would be to use Equation 3.2. Since the mixture contains only iron and sulfur,

$$\% \text{ of iron} + \% \text{ of sulfur} = 100$$

$$\% \text{ of sulfur} = 100 - \% \text{ of iron} = 100 - 21.6 = 78.4$$

Two other useful relations can be obtained from Equation 3.1. We can solve for "amount of *A*," obtaining

$$\text{amount of } A = \text{total amount} \times \frac{\% \text{ of A}}{100} \quad (3.3)$$

Equation 3.3 allows us to calculate the amount of *A* in a sample, knowing its percent and the total amount (Example 3.2a). Alternatively, we can solve for "total amount," obtaining

$$\text{total amount} = \text{amount of } A \times \frac{100}{\% \text{ of } A} \quad (3.4)$$

The use of this equation is shown in Example 3.2b.

Example 3.2 In 1975, 51.2% of the people in the United States were women.

a. In a total population of 214 million, how many were women?
b. How large a group of people would be required to contain 12.0 million women?

Solution

a. Applying Equation 3.3,

$$\text{number of women} = \text{total number} \times \frac{51.2}{100}$$

$$= 214 \text{ million} \times 0.512 = 110 \text{ million}$$

b. Using Equation 3.4,

$$\text{total number} = \text{number of women} \times \frac{100}{51.2}$$

$$= 12.0 \text{ million} \times \frac{100}{51.2} = 23.4 \text{ million}$$

In a group of 23.4 million people, there would be about 12.0 million women (and 11.4 million men).

Percent and Fraction

As we have seen, the composition of a sample can be specified by giving the percents of the different components (A, B, C, . . .). Alternatively, we can cite the decimal fractions of the components:

$$\text{fraction of } A = \frac{\text{amount of } A}{\text{total amount}} \tag{3.5}$$

Comparing Equation 3.5 to Equation 3.1, it should be clear that the percent of a component is 100 times the fraction of that component:

$$\% \text{ of } A = \text{fraction of } A \times 100 \tag{3.6}$$

Thus, if 51.2% of the people in the United States are women, then the fraction of women must be 0.512. If the fraction of passes completed by a quarterback is 0.62, he completes 62% of the passes he throws.

Recall (Equation 3.2) that the percents of all the components of a sample (A, B, C, . . .) must add to 100. In the same way, the fractions must add to unity:

$$\text{fraction of } A + \text{fraction of } B + \text{fraction of } C + \cdots = 1 \tag{3.7}$$

If the fraction of women in the United States is 0.512, the fraction of men must be 1 − 0.512 = 0.488. If the fraction of passes completed by a quarterback is 0.62, then the fraction incomplete or intercepted must be 0.38.

Example 3.3

a. In the mixture referred to in Example 3.1, we found that the mass percents of sulfur and iron were 78.4 and 21.6 in that order. What is the fraction of sulfur in the mixture? the fraction of iron?

b. In an election, candidate A received 12,608 votes, candidate B, 10,192. What fraction of the total vote did A get? what percent?

Solution

a. Applying Equation 3.6,

$$\text{fraction sulfur} = \frac{\%\ \text{sulfur}}{100} = \frac{78.4}{100} = 0.784$$

$$\text{fraction iron} = \frac{\%\ \text{iron}}{100} = \frac{21.6}{100} = 0.216$$

Alternatively, we could have used Equation 3.7 to obtain the fraction of iron:

$$\text{fraction of iron} = 1 - \text{fraction of sulfur} = 1 - 0.784 = 0.216.$$

b. Here, the total vote is 12,608 + 10,192 = 22,800:

$$\text{fraction } A = \frac{12{,}608}{22{,}800} = 0.55298$$

$$\%\ A = 100 \times \text{fraction } A = 100 \times 0.55298 = 55.298$$

3.2 APPLICATIONS IN CHEMISTRY

Many problems in general chemistry are phrased in the language of percent. In this section, we will consider four different areas in which the concept of percent plays an important role.

Concentration of Solutions

The concentration of a solution is often expressed in terms of the percent of a component. We may speak of the **mass percent** of A or its *mole percent*. In either case, the defining equations are entirely analogous to Equation 3.1:

$$\text{mass } \%\ A = \frac{\text{mass of } A}{\text{total mass}} \times 100 \qquad (3.8)$$

$$\text{mole } \%\ A = \frac{\text{no. moles } A}{\text{total no. moles}} \times 100 \qquad (3.9)$$

Ordinarily, the mass percent of a component will differ from its mole percent (Example 3.4).

Example 3.4 A solution is prepared containing 30.0 g of ethyl alcohol, C_2H_5OH, and 30.0 g of water, H_2O.

a. What is the mass percent of ethyl alcohol in this solution?

b. What is the mole percent of ethyl alcohol (1 mol C_2H_5OH weighs 46.07 g; 1 mol H_2O weighs 18.02 g)?

Solution

a. The total mass of the solution is 30.0 g + 30.0 g = 60.0 g. Applying Equation 3.8,

$$\text{mass \% ethyl alcohol} = \frac{30.0\text{ g}}{60.0\text{ g}} \times 100 = 50.0$$

b. Before using Equation 3.9, we must find the number of moles of each component:

$$\text{moles ethyl alcohol} = 30.0\text{ g} \times \frac{1\text{ mol}}{46.07\text{ g}} = 0.651\text{ mol}$$

$$\text{moles water} = 30.0\text{ g} \times \frac{1\text{ mol}}{18.02\text{ g}} = 1.66\text{ mol}$$

Now we apply Equation 3.9:

$$\text{mole \% ethyl alcohol} = \frac{0.651}{0.651 + 1.66} \times 100 = \frac{0.651}{2.31} \times 100 = 28.2$$

Notice that the mole percent of C_2H_5OH is much less than its mass percent. This reflects the fact that a mole of C_2H_5OH weighs much more than a mole of H_2O (46.07 g vs. 18.02 g). Hence, in the same mass, 30.0 g, there are fewer moles of C_2H_5OH than of H_2O.

Equations 3.8 and 3.9 can be solved for the mass or number of moles of a component in solution (compare Equation 3.3):

$$\text{mass of } A = \text{total mass} \times \frac{\text{mass \% } A}{100} \qquad (3.10)$$

$$\text{no. moles } A = \text{total no. moles} \times \frac{\text{mole \% A}}{100} \qquad (3.11)$$

We can also obtain expressions for the total mass or number of moles (compare Equation 3.4):

$$\text{total mass} = \text{mass of } A \times \frac{100}{\text{mass \% } A} \qquad (3.12)$$

$$\text{total no. moles} = \text{no. moles } A \times \frac{100}{\text{mole \% } A} \qquad (3.13)$$

Example 3.5 A certain salt solution contains 3.5% by mass of NaCl. It is desired to recover sodium chloride from the solution by evaporation.

a. What mass of NaCl can be obtained from 62 g of solution?
b. How much solution must be evaporated to obtain 12 g of NaCl?

Solution

a. Using Equation 3.10:

$$\text{mass NaCl} = 62 \text{ g} \times \frac{3.5}{100} = 2.2 \text{ g}$$

b. Using Equation 3.12:

$$\text{total mass} = \text{mass NaCl} \times \frac{100}{\%\ \text{NaCl}} = 12 \text{ g} \times \frac{100}{3.5} = 3.4 \times 10^2 \text{ g}$$

Another concentration unit in common use is **mole fraction:**

$$\text{mole fraction } A = \frac{\text{moles } A}{\text{total no. moles}} \qquad (3.14)$$

We can write equations similar to 3.6 and 3.7 for mole fraction:

$$\text{mole \% } A = \text{mole fraction } A \times 100 \qquad (3.15)$$

$$\text{mole fraction } A + \text{mole fraction } B + \text{mole fraction } C + \cdots = 1 \qquad (3.16)$$

Example 3.6 Dry air consists almost entirely of the three gases N_2, O_2 and Ar. The mole fractions of N_2 and O_2 are 0.781 and 0.210 in that order.

a. What is the mole fraction of Ar?
b. What is the mass in grams of one mole of dry air? (1 mol N_2 = 28.01 g; 1 mol O_2 = 32.00 g; 1 mol Ar = 39.95 g.)

Solution

a. From Equation 3.16,

$$\text{mole fraction Ar} = 1 - \text{mole fraction } N_2 - \text{mole fraction } O_2$$
$$= 1 - 0.781 - 0.210 = 0.009$$

b. In one mole of air there is 0.781 mol N_2, 0.210 mol O_2, and 0.009 mol Ar.

$$\text{mass } N_2 = 0.781 \text{ mol} \times \frac{28.01 \text{ g}}{1 \text{ mol}} = 21.9 \text{ g}$$

$$\text{mass } O_2 = 0.210 \text{ mol} \times \frac{32.00 \text{ g}}{1 \text{ mol}} = 6.7 \text{ g}$$

$$\text{mass Ar} = 0.009 \text{ mol} \times \frac{39.95 \text{ g}}{1 \text{ mol}} = 0.4 \text{ g}$$

$$\text{total mass} = 21.9 \text{ g} + 6.7 \text{ g} + 0.4 \text{ g} = 29.0 \text{ g}$$

We conclude that one mole of air weighs 29.0 g, or that the "molecular mass" of air is 29.0

Isotopic Abundances

Most elements exist in nature as a mixture of isotopes, i.e., atoms of different masses. For example, chlorine occurs as a mixture of two isotopes, Cl-35 and Cl-37, which have atomic masses on the C-12 scale of 34.97 and 36.97 in that order. The relative amounts ("abundances") of the different isotopes of an element are usually specified by giving their mole percents (often referred to simply as "percents"). In the case of chlorine, the mole percents of Cl-35 and Cl-37 are 75.77 and 24.23 in that order.

Knowing the atomic masses and abundances of the isotopes of an element, its atomic mass is readily calculated. The approach used is that followed in Example 3.6b.

Example 3.7 From the information just given, calculate the atomic mass of chlorine.

Solution

We are asked in effect to calculate the mass in grams of one mole of Cl. To do this, we proceed as in Example 3.6b, first writing down the mole fractions of the two isotopes.

$$\text{mole fraction Cl-35} = 75.77/100 = 0.7577$$

$$\text{mole fraction Cl-37} = 24.23/100 = 0.2423$$

In one mole of Cl, there is 0.7577 mol of Cl-35 (molar mass 34.97 g) and 0.2423 mol of Cl-37 (molar mass 36.97 g). The mass of one mole of Cl must then be

$$0.7577 \times 34.97 \text{ g} + 0.2423 \times 36.97 \text{ g} = 26.49 \text{ g} + 8.96 \text{ g} = 35.45 \text{ g}$$

The atomic mass of chlorine is 35.45.

The calculation that we have just gone through can be turned around. For an element consisting of only two isotopes, we can calculate the relative abundances of these isotopes, knowing the atomic mass of the element (Example 3.8).

Example 3.8 Boron consists of two isotopes: B-10, atomic mass 10.013, and B-11, atomic mass 11.009. The atomic mass of the element itself is 10.811. What are the mole fractions and percents of the two isotopes?

Solution

The basic relation (using the abbreviation A.M. for atomic mass) is

$$\text{A.M. B} = \text{mole fraction B-10 (A.M. B-10)} + \text{mole fraction B-11 (A.M. B-11)}$$

Substituting the atomic masses given in the statement of the problem,

$$10.811 = (\text{mole fract. B-10}) \times 10.013 + (\text{mole fract. B-11}) \times 11.009$$

To solve this equation, we must relate the mole fractions of the two isotopes to each other. This is readily done: *since there are only two isotopes, their mole fractions must add up to 1.* Thus, if we let x be the mole fraction of B-11, the mole fraction of B-10 must be $1 - x$. Making this substitution,

$$10.811 = 10.013(1 - x) + 11.009\,x$$

Solving this equation for x,

$$10.811 - 10.013 = 11.009\,x - 10.013\,x; \quad 0.798 = 0.996\,x$$

$$x = 0.798/0.996 = 0.801$$

We conclude that the mole fraction of B-11 must be 0.801. That of B-10 must be $1 - 0.801 = 0.199$. The mole percents are 80.1 for B-11 and 19.9 for B-10.

Percent Composition: Formulas of Compounds

The composition of a compound is most often specified by giving its formula. The *simplest formula* of a compound gives us the simplest atom ratio or mole ratio of the elements present. Consider, for example, the simplest formula of calcium chloride, $CaCl_2$. In this compound, the atom ratio or the mole ratio of calcium to chlorine is 1 : 2. There are two chlorine atoms for every calcium atom; there are 2 mol of Cl (70.90 g) for every mol of Ca (40.08 g).

From the formula of a compound, the mass percents of the elements present are readily determined (Example 3.9).

Example 3.9 Determine the mass percents of the elements in potassium chromate, which has the simplest formula K_2CrO_4 (A.M. K = 39.1, Cr = 52.0, O = 16.0).

Solution

Let us base our calculations on one mole of K_2CrO_4, which contains

2 mol K, 1 mol Cr, 4 mol O

Since the atomic mass of potassium is 39.1, one mole of K weighs 39.1 g. Similarly, one mole of Cr weighs 52.0 g; one mole of O weighs 16.0 g. Hence, in one mole of K_2CrO_4 we have

$$\begin{aligned} 2 \times 39.1\text{ g} &= 78.2\text{ g K} \\ 1 \times 52.0\text{ g} &= 52.0\text{ g Cr} \\ 4 \times 16.0\text{ g} &= \underline{64.0\text{ g O}} \\ & 194.2\text{ g } K_2CrO_4 \end{aligned}$$

The mass percents of the three elements must then be

$$\text{mass \% K} = \frac{78.2}{194.2} \times 100 = 40.3; \qquad \text{mass \% Cr} = \frac{52.0}{194.2} \times 100 = 26.8$$

$$\text{mass \% O} = 100 - 40.3 - 26.8 = 32.9$$

We have just seen how the simplest formula of a compound can be used to obtain the mass percents of the elements present. It is also possible to carry out the reverse calculation. That is, knowing the mass percents, we can deduce the simplest formula. The reasoning in this case is a bit more subtle and is illustrated in some detail in Example 3.10.

Example 3.10 The mass percents of calcium, chlorine and oxygen in calcium chlorite are 22.9, 40.5, and 36.6 in that order. What is the simplest formula of calcium chlorite? (A.M. Ca = 40.1, Cl = 35.5, 0 = 16.0.)

Solution

We are looking for the atom ratio or the mole ratio (they are identical) of Ca to Cl to O in this compound. A logical approach would be to first find the number of moles of each element in a fixed mass of the compound which, for convenience in calculations, we take to be 100 g. In 100 g of the compound, there are 22.9 g of Ca, 40.5 g of Cl, and 36.6 g of O. Converting to moles,

$$\text{moles Ca} = 22.9 \text{ g Ca} \times \frac{1 \text{ mol Ca}}{40.1 \text{ g Ca}} = 0.571 \text{ mol Ca}$$

$$\text{moles Cl} = 40.5 \text{ g Cl} \times \frac{1 \text{ mol Cl}}{35.5 \text{ g Cl}} = 1.14 \text{ mol Cl}$$

$$\text{moles O} = 36.6 \text{ g O} \times \frac{1 \text{ mol O}}{16.0 \text{ g O}} = 2.29 \text{ mol O}$$

The numbers we have just calculated, 0.571, 1.14 and 2.29, represent the relative numbers of moles of Ca, Cl, and O in calcium chlorite. To find the formula, we need the *simplest* mole ratio of Ca : Cl : O. One way to obtain this ratio is to divide each of these numbers by the smallest, 0.571:

$$\text{Ca: } \frac{0.571}{0.571} = 1.00 \qquad \text{Cl: } \frac{1.14}{0.571} = 2.00 \qquad \text{O: } \frac{2.29}{0.571} = 4.00$$

This calculation tells us that for every mole of Ca, we have two moles of Cl and four moles of O. The simplest formula of calcium chlorite must then be $CaCl_2O_4$.

Sometimes, in a calculation of this sort, the "whole number ratio" arrived at in the last step is not too obvious. Consider, for example, the gaseous hydrocarbon known as propane. It turns out that in 100 g of propane there are

6.80 mol C and 18.14 mol H

Dividing each number by the smallest, 6.80 we obtain

$$\text{C: } \frac{6.80}{6.80} = 1.00 \qquad \text{H: } \frac{18.14}{6.80} = 2.67$$

This tells us that the atom ratio of H to C is

$$\frac{2.67}{1.00}$$

To obtain the simplest whole number ratio, we multiply both numerator and denominator by 3:

$$\frac{2.67 \times 3}{1.00 \times 3} = \frac{8.01}{3.00} = \frac{8}{3}$$

We conclude that the simplest formula of propane is C_3H_8. In general, if the atom ratio is not itself a whole number (e.g., 1.50, 2.67, . . .), you can multiply numerator and denominator by 2 or 3 or . . . to obtain a ratio of two integers (e.g., 3/2, 8/3, . . .).

Percent Yield

The percent yield of a product, *P*, in a reaction is defined as

$$\% \text{ yield of } P = \frac{\text{actual yield of } P}{\text{theoretical yield of } P} \times 100 \qquad (3.17)$$

The theoretical yield is the maximum amount of product that could be obtained, based on the amounts of reactants used. In a given reaction, the percent yield can vary from 0 to 100, depending upon a variety of factors.

Example 3.11 Consider the reaction of toluene with nitric acid to give trinitrotoluene (TNT):

$$\underset{\text{toluene}}{C_7H_8\ (l)} + \underset{\text{nitric acid}}{3\ HNO_3\ (l)} \rightarrow \underset{\text{TNT}}{C_7H_5(NO_2)_3\ (s)} + 3\ H_2O\ (l)$$

Suppose 20.0 g of toluene is added to an excess of nitric acid.

a. Using the method described in Example 2.6c, p. 23, determine the mass of TNT that can be produced from 20.0 g of toluene. This is the theoretical yield of TNT in this reaction.
b. If 21.5 g of TNT is actually obtained, what is the percent yield?
c. If the percent yield were 62.0, what mass of TNT would be produced?

Solution

a. We follow a three-step path:
   1. grams toluene to moles toluene (1 mol C_7H_8 = 92.0 g)
   2. moles toluene to moles TNT (1 mol $C_7H_8 \simeq$ 1 mol TNT)
   3. moles TNT to grams TNT (1 mol $C_7H_5(NO_2)_3$ = 227 g)

$$\text{theor. yield TNT} = 20.0\text{ g } C_7H_8 \times \underset{(1)}{\frac{1\text{ mol } C_7H_8}{92.0\text{ g } C_7H_8}} \times \underset{(2)}{\frac{1\text{ mol TNT}}{1\text{ mol } C_7H_8}} \times \underset{(3)}{\frac{227\text{ g TNT}}{1\text{ mol TNT}}} = 49.3\text{ g TNT}$$

b. $\% \text{ yield TNT} = \frac{\text{actual yield TNT}}{\text{theor. yield TNT}} \times 100 = \frac{21.5\text{ g}}{49.3\text{ g}} \times 100 = 43.6$

c. $\text{actual yield TNT} = \text{theor. yield TNT} \times \frac{\%\text{ yield TNT}}{100}$

$$= 49.3\text{ g} \times \frac{62.0}{100} = 30.6\text{ g}$$

PROBLEMS

3.1 A box contains 50 red marbles and 42 black marbles.

a. What is the percent of red marbles? black marbles?

b. What is the decimal fraction of red marbles? black marbles?

c. Suppose each red marble weighs 11 g and each black marble weighs 13 g. What is the total mass of the marbles? What is the mass percent of red marbles?

3.2 About 10% of the population are left-handed.

a. In a group of 250 people, how many would you expect to be left-handed?

b. How large a group of people would be required to contain 120 left-handed people?

3.3 A certain brand of aspirin tablets is advertised as being 98% pure.

a. How many grams of aspirin can be obtained from 1.00 kg of tablets?

b. How many grams of tablets must be taken to get 1.00 kg of aspirin?

3.4 A person has a pile of coins consisting of 50 nickels (5¢) and 20 dimes (10¢).

a. What is the percent of nickels? dimes?

b. What is the total value, in cents, of the coins?

c. What is the average value of one of the 70 coins in this pile?

3.5 A mixture of 80 coins (dimes and nickels) has a total value of $5.

a. Set up an equation for the total value in terms of the number of dimes, x, and the number of nickels, $80-x$.

b. Solve the equation in (a) to find the number of dimes present; the number of nickels.

c. What is the percent of dimes? the fraction of dimes?

3.6 A basket of fruit contains 4 apples, 3 oranges, and 1 grapefruit. The apples weigh 40 g apiece, the oranges 40 g, and the grapefruit 120 g.

a. What is the total mass of the fruit in the basket?

b. What percent of this mass is accounted for by the apples? the oranges? the grapefruit?

3.11 A solution contains 1.0 mol of methyl alcohol and 2.5 mol of water.

a. What is the mole percent of methyl alcohol? water?

b. What is the mole fraction of methyl alcohol? water?

c. One mole of methyl alcohol, CH_3OH, weighs 32.0 g; one mole of water, H_2O, weighs 18.0 g. What is the total mass of the solution? What is the mass percent of methyl alcohol?

3.12 Iron(III) oxide, Fe_2O_3, contains 69.9% by mass of iron.

a. How much iron can be obtained from 250 g of iron(III) oxide?

b. How much iron(III) oxide would be required to contain 120 g of iron?

3.13 The mole percent of N_2 in air is 78.08.

a. How many moles of N_2 are present in 1.20 mol of air?

b. How many moles of air must be taken to obtain 1.00 mol of N_2?

3.14 Gallium consists of two isotopes with atomic masses of 68.93 and 70.93. The mass percents of these two isotopes, in the same order, are 60.16 and 39.84. What is the atomic mass of gallium?

3.15 The atomic mass of copper is 63.54. It consists of two isotopes, Cu-63 and Cu-65, with atomic masses of 62.96 and 64.96 in that order. Calculate the mole percents of the two isotopes.

3.16 The simplest formula of calcium chlorate is $CaCl_2O_6$. Taking the atomic masses of Ca, Cl, and O to be 40.1, 35.5, and 16.0 in that order, determine

a. the mass of one mole of $CaCl_2O_6$.

b. the mass percents of calcium, chlorine, and oxygen in calcium chlorate.

3.7 A candy-store owner prepares a mixture of jelly beans and gum drops, adding 3 parts by mass (e.g., 3.0 kg) of jelly beans to 2 parts by mass (e.g., 2.0 kg) of gum drops.

a. What is the mass percent of jelly beans in the mixture? the mass percent of gum drops?

b. In 100 g of the mixture, what is the mass of jelly beans? of gum drops?

c. Suppose a jelly bean weighs 3.0 g and a gum drop weighs 1.0 g. How many jelly beans are there in 100 g of the mixture? how many gum drops?

d. What is the ratio of gum drops to jelly beans in the mixture?

3.8 A mixture consists of 82 McIntosh, 164 Cortland, and 205 Delicious apples.

a. Per McIntosh, how many Cortlands are there? how many Delicious?

b. What is the simplest, whole-number ratio of McIntosh : Cortland : Delicious?

3.9 A mixture of sand and salt, used on roads in winter, contains 35% by mass of salt.

a. What is the mass percent of sand?

b. What is the mass ratio of salt to sand?

3.10 At best, a Maine farmer can expect to get 20 bushels of potatoes for every bushel he plants.

a. What is the "theoretical yield" the farmer can hope for if he plants 2.5 bushels?

b. If he actually gets 32 bushels, what is his percent yield?

c. Based on the percent yield in (b), how many bushels of potatoes should the farmer plant if he needs to harvest 60 bushels?

3.17 A certain compound contains 19.3% Na, 26.8% S, and 53.9% O by mass.

a. What are the masses of Na, S, and O in a 100-g sample of the compound?

b. The molar masses of Na, S, and O are 23.0 g, 32.1 g, and 16.0 g in that order. Find the numbers of moles of Na, S, and O in the 100-g sample.

c. What is the mole ratio of S to Na? of O to Na?

d. What is the simplest formula of the compound?

3.18 In 100 g of benzoic acid, there are 5.74 mol C, 4.92 mol H, and 1.64 mol O.

a. Per mole of oxygen, how many moles of hydrogen are there? how many moles of carbon?

b. What is the simplest formula of benzoic acid?

3.19 Hemoglobin forms a complex with oxygen, $Hem \cdot O_2$, and another with carbon monoxide, $Hem \cdot CO$. When 20% of the hemoglobin is tied up as the CO complex, death results. Under these conditions,

a. What is the percent of $Hem \cdot O_2$?

b. What is the ratio of $Hem \cdot O_2/Hem \cdot CO$?

3.20 Consider the reaction

$$CO\ (g) + 2\ H_2\ (g) \rightarrow CH_3OH\ (l)$$

At best, in this reaction a chemist can get 1.14 g of methyl alcohol, CH_3OH, for every gram of carbon monoxide, CO, that she starts with.

a. What is the theoretical yield of CH_3OH, starting with 2.50 g of CO?

b. If the actual yield of CH_3OH is 2.24 g, what is the percent yield?

c. Based on the percent yield in (b), how many grams of CO should she start with to produce 2.85 g of CH_3OH?

CHAPTER 4

Exponential Numbers

In chemistry, we frequently deal with very large or very small numbers. In one gram of the element carbon there are

$$50{,}150{,}000{,}000{,}000{,}000{,}000{,}000$$

atoms of carbon. At the opposite extreme, the mass of a single carbon atom is

$$0.000\ 000\ 000\ 000\ 000\ 000\ 000\ 019\ 94\ \text{g}$$

Numbers such as these are not only difficult to write, they are very awkward to work with. For example, neither of the numbers just written could be entered directly into a calculator. To simplify operations involving very large or very small numbers, we use what is known as **exponential** or *scientific* notation. In exponential notation, numbers such as those written above are expressed as a number between one and ten (**coefficient**) times an integral power of ten (**exponential**). Examples of exponential numbers include

$$1 \times 10^4 \qquad 2.23 \times 10^3 \qquad 5.6 \times 10^{-4}$$

To understand precisely what exponential numbers mean, it is helpful to refer to Table 4.1. For the three numbers just cited we have

$$1 \times 10^4 = 1(10{,}000) = 10{,}000$$

$$2.23 \times 10^3 = 2.23(1000) = 2230$$

$$5.6 \times 10^{-4} = 5.6(0.0001) = 0.00056$$

Table 4.1 EXPONENTIALS

10^6 =	(10)(10)(10)(10)(10)(10)	=	1,000,000
10^5 =	(10)(10)(10)(10)(10)	=	100,000
10^4 =	(10)(10)(10)(10)	=	10,000
10^3 =	(10)(10)(10)	=	1000
10^2 =	(10)(10)	=	100
10^1 =	(10)	=	10
10^0 =		=	1
10^{-1} =	(0.1)	=	0.1
10^{-2} =	(0.1)(0.1)	=	0.01
10^{-3} =	(0.1)(0.1)(0.1)	=	0.001
10^{-4} =	(0.1)(0.1)(0.1)(0.1)	=	0.0001

4.1 WRITING NUMBERS IN EXPONENTIAL NOTATION

To make use of exponential notation, we must be able to write any number, large or small, as an exponential number. To understand how this is done, it may be helpful to start with two simple cases which can be worked directly from the entries in Table 4.1.

Suppose we wish to express the number 5196 in exponential notation. We realize that this number can be written as

$$5.196 \times 1000$$

Referring to Table 4.1, we note that $1000 = 10^3$. Therefore,

$$5196 = 5.196 \times 1000 = 5.196 \times 10^3$$

As another illustration, consider the number 0.0028. To express this number in exponential notation, we first write it as

$$2.8 \times 0.001$$

Since $0.001 = 10^{-3}$, we have

$$0.0028 = 2.8 \times 0.001 = 2.8 \times 10^{-3}$$

The method just used is not very useful with extremely large or extremely small numbers, for which tables of exponentials are seldom available. We can, however, deduce from these examples a more general approach to the problem. Notice that when we expressed 5196 in exponential notation, we arrived at an exponent of 3. *This is the number of places that the decimal point must be moved (to the left) to give the coefficient, 5.196.* Again, in expressing 0.0028 in exponential notation, the exponent, −3, tells us *the number of places that the decimal point must be moved (to the right) to give the coefficient, 2.8.* In general:

To express a number in exponential notation, write it in the form

$$C \times 10^n$$

where C is a number between 1 and 10 (e.g., 1, 2.62, 5.8) **and n is a positive or negative integer** (e.g., 1, −1, −3). **To find n, count the number of places that the decimal point must be moved to give the coefficient C. If the decimal point must be moved to the left, n is a positive integer. If it must be moved to the right, n is a negative integer.**

Two simple examples illustrate this rule:

$0.000569 = 5.69 \times 10^{-4}$ (decimal point must be moved 4 places to the right)
$21982 = 2.1982 \times 10^4$ (decimal point must be moved 4 places to the left)

Example 4.1 Express the two numbers given at the beginning of the chapter (the number of atoms in one gram of carbon and the mass in grams of one carbon atom) in exponential notation.

Solution

For the number

$$50{,}150{,}000{,}000{,}000{,}000{,}000{,}000$$

the coefficient is 5.015. To obtain this coefficient, the decimal point must be moved 22 places (count them!) to the *left*. It follows that the exponential number is

$$5.015 \times 10^{22}$$

Similarly, the coefficient of the number

$$0.000\ 000\ 000\ 000\ 000\ 000\ 000\ 019\ 94$$

is 1.994. The decimal point must be moved 23 places to the *right* to obtain the coefficient. Therefore, we obtain

$$1.994 \times 10^{-23}$$

4.2 MAGNITUDE OF EXPONENTIAL NUMBERS

The magnitude of a number written in exponential notation depends upon the value of both the coefficient, C, and the exponent, n. Suppose we compare two numbers which have the same value of n, such as 2.6×10^2 and 3.8×10^2. Here, *the larger number is the one with the larger coefficient,* as you can readily see by converting to ordinary numbers:

$$3.8 \times 10^2 > 2.6 \times 10^2; \qquad 380 > 260$$

This rule holds regardless of what the exponent is, provided it is the same for both numbers:

$$3.8 \times 10^{-3} > 2.6 \times 10^{-3}; \qquad 0.0038 > 0.0026$$

More commonly, we want to compare the magnitudes of exponential numbers with different values of n. Here, **the larger number is the one which has the algebraically larger value of n.** This means that

$$2.6 \times 10^2 > 4.8 \times 10^1; \qquad 260 > 48$$
$$3.2 \times 10^1 > 8.0 \times 10^{-1}; \qquad 32 > 0.80$$
$$2 \times 10^{-2} > 4 \times 10^{-3}; \qquad 0.02 > 0.004$$

Notice that it is the *algebraic* value of the exponent that is important. Since -2 is algebraically larger than -3, 2×10^{-2} is a larger number than 4×10^{-3}. In the same way, 10^{-15} is larger than 10^{-16}, and so on.

Example 4.2 Arrange the following numbers in order of decreasing magnitude, starting with the largest:

$$6.0 \times 10^{-22}, \quad 3.0 \times 10^{4}, \quad 4.1 \times 10^{-14}, \quad 5.5 \times 10^{-22}, \quad 3.8 \times 10^{-2}$$

Solution

We first apply the rule that the larger of two numbers is the one with the larger exponent. This tells us that $3.0 \times 10^4 > 3.8 \times 10^{-2}$, $3.8 \times 10^{-2} > 4.1 \times 10^{-14}$, and so on. That is, on the basis of this rule alone, we can say that

$$3.0 \times 10^4 > 3.8 \times 10^{-2} > 4.1 \times 10^{-14} > (\text{both } 5.5 \times 10^{-22} \text{ and } 6.0 \times 10^{-22})$$

Then we apply the rule that, for two numbers with the same value of n, the larger is the one with the larger coefficient. This tells us that $6.0 \times 10^{-22} > 5.5 \times 10^{-22}$. The complete sequence is

$$3.0 \times 10^4 > 3.8 \times 10^{-2} > 4.1 \times 10^{-14} > 6.0 \times 10^{-22} > 5.5 \times 10^{-22}$$

4.3 MULTIPLICATION AND DIVISION

A major advantage of exponential notation is that it greatly simplifies the processes of multiplication and division. In applying these processes to exponential numbers, we use the fact that *to multiply, we add exponents:*

$$10^1 \times 10^2 = 10^{1+2} = 10^3$$

$$10^6 \times 10^{-4} = 10^{6+(-4)} = 10^2$$

To divide, we subtract exponents:

$$10^3/10^2 = 10^{3-2} = 10^1$$

$$10^{-3}/10^6 = 10^{-3-6} = 10^{-9}$$

$$10^4/10^{-6} = 10^{4-(-6)} = 10^{10}$$

Using these principles, we arrive at the following rules for multiplying or dividing exponential numbers:

To multiply one exponential number by another, first multiply the coefficients together in the usual way. Then add exponents.

To divide one exponential number by another, divide coefficients in the usual way and then subtract exponents.

Example 4.3 Carry out the indicated operations:

a. $(5.00 \times 10^4) \times (1.60 \times 10^2)$
b. $(6.01 \times 10^{-3})/(5.23 \times 10^6)$

Solution

a. For convenience, we first separate the coefficients from the exponential terms:

$$(5.00 \times 1.60) \times (10^4 \times 10^2)$$

Multiplying coefficients and adding exponents, we obtain 8.00×10^6. (Here and throughout this chapter, we use the rules discussed in Chapter 6 to express answers to the correct number of significant figures.)

b. $(6.01 \times 10^{-3})/(5.23 \times 10^6) = \dfrac{6.01}{5.23} \times \dfrac{10^{-3}}{10^6} = 1.15 \times 10^{-9}$

Often, when exponential numbers are multiplied or divided, the answer is not in standard exponential notation. Consider, for example,

$$(5.0 \times 10^4) \times (6.0 \times 10^3)$$

Multiplying in the usual way, we obtain

$$(5.0 \times 6.0) \times (10^4 \times 10^3) = 30 \times 10^7$$

Again,

$$(3.60 \times 10^2)/(4.92 \times 10^4) = \frac{3.60}{4.92} \times \frac{10^2}{10^4} = 0.732 \times 10^{-2}$$

The two numbers just obtained, 30×10^7 and 0.732×10^{-2}, are not in standard exponential notation. The coefficients are *not* numbers between 1 and 10. To express these numbers in exponential notation, we proceed as indicated in Example 4.4.

Example 4.4 Express the numbers 30×10^7 and 0.732×10^{-2} in standard exponential notation. That is, write them as a number between 1 and 10, times 10 to the proper power.

Solution

In the first case, we could write

$$30 \times 10^7 = (3.0 \times 10^1) \times 10^7 = 3.0 \times 10^8$$

What we did here was to first write the number 30 in exponential notation. To do this, we followed the rule cited on p. 41. Since the decimal point had to be moved one place to the *left* to change 30 to 3.0, the exponent was *+1* (i.e., $30 = 3.0 \times 10^1$). Then, we multiplied 3.0×10^1 by 10^7. Adding exponents, we obtained 3.0×10^8.

We follow a similar procedure in the second case:

$$0.732 \times 10^{-2} = (7.32 \times 10^{-1}) \times 10^{-2} = 7.32 \times 10^{-3}$$

Again, we started by expressing the coefficient, 0.732, in exponential notation. Since the decimal point had to be moved one place to the *right* to change 0.732 to 7.32, the exponent was -1 (i.e., $0.732 = 7.32 \times 10^{-1}$). Then we multiplied 7.32×10^{-1} by 10^{-2}, again adding exponents to obtain 7.32×10^{-3}.

In these cases, and in all others of a similar type, we first *express the coefficient in standard exponential notation* and go on from there, using the rules governing the use of exponents.

4.4 RAISING TO POWERS AND EXTRACTING ROOTS

To raise an exponential number to a power, we use the fact that

$$\mathbf{(10^a)^b = 10^{a \times b}}$$

To illustrate this rule, consider:

$$(10^2)^3 = 10^2 \times 10^2 \times 10^2 = 10^6 = 10^{(2 \times 3)}$$

$$(10^{-2})^4 = 10^{-2} \times 10^{-2} \times 10^{-2} \times 10^{-2} = 10^{-8} = 10^{(-2 \times 4)}$$

To raise an exponential number to a power, we treat the coefficient in the usual manner and use the rule just given to find the value of the exponent:

$$(2.0 \times 10^{-3})^2 = (2.0)^2 \times (10^{-3})^2 = 4.0 \times 10^{-6}$$

$$(3.0 \times 10^2)^3 = (3.0)^3 \times (10^2)^3 = 27 \times 10^6 = 2.7 \times 10^7$$

The same principal can be used to extract a root (square root, cube root, etc.) of an exponential number. Here we are dealing with a fractional power:

$$\sqrt[2]{10} = 10^{1/2}; \qquad \sqrt[3]{10} = 10^{1/3}; \qquad \sqrt[n]{10} = 10^{1/n}$$

but the operation is entirely analogous. Thus,

$$\sqrt[2]{10^6} = (10^6)^{1/2} = 10^{(6/2)} = 10^3$$

$$\sqrt[2]{4.0 \times 10^6} = (4.0 \times 10^6)^{1/2} = (4.0)^{1/2} \times (10^6)^{1/2} = 2.0 \times 10^3$$

As before, we operate on the coefficient and exponential separately.

Extracting square roots can pose a problem when the exponent is not an even number, i.e., divisible by 2 to give an integer. Consider, for example,

$$(4.0 \times 10^5)^{1/2}$$

If we follow the procedure described above, we obtain

$$(4.0)^{1/2} \times (10^5)^{1/2} = 2.0 \times 10^{5/2}$$

The answer is not in standard exponential form. Indeed, $10^{5/2}$ is an extremely awkward expression to work with because it is not readily translated into an ordinary number.

One way to handle cases of this type* is to rewrite the exponential number so as to make the exponent, n, an even number. To do this here, we divide the exponential term by 10 and multiply the coefficient by 10:

$$(4.0 \times 10^5)^{1/2} = (40 \times 10^4)^{1/2}$$

Now, proceeding in the usual manner, we obtain

$$(40 \times 10^4)^{1/2} = (40)^{1/2} \times (10^4)^{1/2} = 6.3 \times 10^2$$

(The square root of 40 is found on a calculator to be 6.3.)

From this example, we can draw a general rule. To extract the square root of an exponential number where n is odd (e.g., -1, 3, 5), we convert n to an even number by dividing the exponential term by 10. At the same time, we multiply the coefficient by 10 to keep the value unchanged. Thus we have

$$(1.0 \times 10^{-1})^{1/2} = (10 \times 10^{-2})^{1/2} = 3.2 \times 10^{-1}$$

$$(5.0 \times 10^3)^{1/2} = (50 \times 10^2)^{1/2} = 7.1 \times 10^1$$

The same principle can be used for higher roots. To extract the nth root of an exponential number, we make sure that the exponent, when divided by n, gives a whole number. Consider, for example, the cube root of 2.0×10^7. We rewrite this as 20×10^6 (multiplying 2.0 by 10, dividing 10^7 by 10). Then we take a cube root in the usual manner:

$$(2.0 \times 10^7)^{1/3} = (20 \times 10^6)^{1/3} = 20^{1/3} \times 10^2 = 2.7 \times 10^2$$

Example 4.5 Perform the indicated operations:

a. $(6.2 \times 10^{-4})^2$ b. $(3.0 \times 10^6)^{1/2}$ c. $(2.81 \times 10^{-5})^{1/2}$

Solution

a. $(6.2 \times 10^{-4})^2 = (6.2)^2 \times 10^{-8} = 38 \times 10^{-8} = (3.8 \times 10^1) \times 10^{-8} = 3.8 \times 10^{-7}$

b. $(3.0 \times 10^6)^{1/2} = (3.0)^{1/2} \times 10^3 = 1.7 \times 10^3$

c. Here we first operate on the number so that the exponent will give an integer when divided by 2. To do this, we divide the exponential term by 10 and multiply the coefficient by 10.

$$(2.81 \times 10^{-5})^{1/2} = (28.1 \times 10^{-6})^{1/2} = (28.1)^{1/2} \times (10^{-6})^{1/2} = 5.30 \times 10^{-3}$$

---

*Another approach is to enter the number on your calculator in exponential notation and then take the square root (p. 49).

4.5 ADDITION AND SUBTRACTION

Sometimes we find it necessary to add or subtract two exponential numbers. These processes are simple if both exponents are the same. To add

$$2.02 \times 10^7 + 3.16 \times 10^7$$

we factor to obtain

$$(2.02 + 3.16) \times 10^7 = 5.18 \times 10^7$$

In the same way

$$6.1 \times 10^{-5} - 3.0 \times 10^{-5} = (6.1 - 3.0) \times 10^{-5} = 3.1 \times 10^{-5}$$

If the exponents are different, the numbers can be operated on to make the exponents the same. This procedure is illustrated in Example 4.6.

Example 4.6 Carry out the indicated operations:

a. $6.04 \times 10^3 + 2.6 \times 10^2$ b. $9.82 \times 10^{-4} - 8.2 \times 10^{-5}$

Solution

a. We cannot perform the addition directly, any more than we can add six oranges to two apples. To add, we must make the exponents of the two numbers the same. One way to do this is to operate on the second number, expressing it as a coefficient times 10^3:

$$2.6 \times 10^2 = 0.26 \times 10^3$$

Now we can add:

$$6.04 \times 10^3 + 0.26 \times 10^3 = 6.30 \times 10^3$$

b. Proceeding as in part a

$$9.82 \times 10^{-4} - 8.2 \times 10^{-5} = 9.82 \times 10^{-4} - 0.82 \times 10^{-4} = 9.00 \times 10^{-4}$$

Alternatively, we could have operated on the first number rather than the second:

$$9.82 \times 10^{-4} = 98.2 \times 10^{-5}$$

Hence,

$$9.82 \times 10^{-4} - 8.2 \times 10^{-5} = 98.2 \times 10^{-5} - 8.2 \times 10^{-5} = 90.0 \times 10^{-5} = 9.00 \times 10^{-4}$$

This procedure gives the same answer, of course. Notice, though, that one extra step is involved, since the answer obtained directly, 90.0×10^{-5}, is not in standard exponential notation. For that reason, it is a bit more efficient to operate on the smaller number (8.2×10^{-5}) rather than the larger number (9.82×10^{-4}).

From this example, we can draw a general rule for addition or subtraction of two exponential numbers where the exponents differ. One of the numbers must be operated upon to make the exponents the same. It is simplest to work with the smaller exponential number (the one for which n is algebraically smaller). The exponent of that term is multiplied by the appropriate power of 10; the coefficient is divided by the same power of 10.

4.6 EXPONENTIAL NOTATION ON THE CALCULATOR

All the operations discussed in Sections 4.3–4.5, at least in principle, can be carried out on an electronic calculator. To do this, you must learn how to enter numbers in exponential notation. This procedure involves the use of an "exponential notation" key, most often marked " EE↓ " or " EEX ." To illustrate, consider how you would enter 1.2×10^4:

Press	Display		
1.2	**1.2**		
EE↓	**1.2**	**00**	
4	**1.2**	**04**	(read as 1.2×10^4)

One more step is involved if the exponent is negative. The "change sign" key, marked +/− or CHS must be used. Thus, to enter 2.0×10^{-6}

Press	Display		
2.0	**2.0**		
EE↓	**2.0**	**00**	
6	**2.0**	**06**	
+/−	**2.0**	**−06**	(read as 2.0×10^{-6})

Your calculator will carry out arithmetical operations with exponential numbers. The procedure is similar to that described in Chapter 1 for ordinary numbers. In most cases, though, it is quite tedious. Suppose, for example, you want to multiply 1.2×10^4 by 2.0×10^{-6}. Using exponential notation on your calculator, nine steps are required. On an "algebraic" calculator (Chapter 1), they are:

Press	Display		
1.2	**1.2**		
EE↓	**1.2**	**00**	
4	**1.2**	**04**	
×	**1.2**	**04**	
2.0	**2.0**		
EE↓	**2.0**	**00**	
6	**2.0**	**06**	
+/−	**2.0**	**−06**	
=	**2.4**	**−02**	(read as 2.4×10^{-2})

Here it would be simpler to follow the rules for multiplication of exponential numbers described in Section 4.3. Realizing that $10^4 \times 10^{-6} = 10^{-2}$, eliminates five of the nine steps!

$$(1.2 \times 10^4) \times (2.0 \times 10^{-6}) = (1.2 \times 2.0) \times 10^{4-6} = 2.4 \times 10^{-2}$$

The example just cited is typical of most calculations involving exponential numbers. You save time and avoid confusion by learning the rules discussed in Sections 4.3-4.5 and applying them to the problem at hand. In this way you reduce the number of steps required with your calculator. There are, however, a few operations where exponential notation on your calculator can be used to advantage. These include:

1. Adding or subtracting exponential numbers where the exponents differ. The "exponential shift" described in Example 4.6 is unnecessary if you use your calculator in exponential notation.
2. Taking the square root of an exponential number where n is odd ($n = -1, 3, 5, \ldots$). Here again, by using your calculator in exponential notation, you can obtain an answer directly. You do not have to first change the form of the number as described in Example 4.5, part c.

Example 4.7 Using your calculator in exponential notation, evaluate:

a. $6.04 \times 10^3 + 2.6 \times 10^2$
b. $(2.81 \times 10^{-5})^{1/2}$

Solution

The steps involved with a typical "algebraic" calculator are:

(a)

Press	Display	
6.04	**6.04**	
EE↓	**6.04**	**00**
3	**6.04**	**03**
+	**6.04**	**03**
2.6	**2.6**	
EE↓	**2.6**	**00**
2	**2.6**	**02**
=	**6.30**	**03**

(b)

Press	Display	
2.81	**2.81**	
EE↓	**2.81**	**00**
5	**2.81**	**05**
+/−	**2.81**	**−05**
$\sqrt{x}$	**5.30**	**−03**

The answers are, of course, the same as those obtained earlier in Examples 4.5 and 4.6:

$$6.04 \times 10^3 + 2.6 \times 10^2 = 6.04 \times 10^3 + 0.26 \times 10^3 = 6.30 \times 10^3$$

$$(2.81 \times 10^{-5})^{1/2} = (28.1 \times 10^{-6})^{1/2} = 5.30 \times 10^{-3}$$

PROBLEMS

Here, as in all chapters, the problems are arranged in matched pairs. The problem at the left illustrates a simple mathematical principle applied to exponential numbers. The "word problem" at the right illustrates the same principle as it applies in general chemistry. Don't panic if some of the words used are unfamiliar to you. If you read the problem carefully and apply the rules for working with exponents, you should have no trouble answering it. Indeed, you may learn a surprising amount of chemistry in the process.

4.1 Express the following numbers in exponential notation:

a. 1000
b. one billion
c. 0.000 001
d. 16,220
e. 212.6
f. 0.189
g. 6.18
h. 0.000 000 078 46

4.11 A helium atom has the following properties:

Mass = 0.000 000 000 000 000 000 000 006 65 g

radius = 0.000 000 004 6 cm

average speed at 25° C = 136,000 cm/s

Express these quantities in standard exponential notation.

4.2 Express as ordinary numbers:

a. 1.2×10^{-3}
b. 6.4×10^{1}
c. 3.0×10^{0}
d. 4.1×10^{5}
e. 1.4×10^{-6}

4.12 Complete the following table, giving the solubilities of several different compounds as ordinary numbers (first column) or exponential numbers (second column).

AgCl	______	=	1.3×10^{-5}
$PbCl_2$	0.016	=	______
$AgC_2H_3O_2$	0.045	=	______

4.3 In each of the following pairs, select the number which is larger:

a. 3×10^{3}; 3×10^{-3}
b. 3×10^{3}; 10,000
c. 0.0001; 2×10^{-4}
d. 6×10^{7}; 4×10^{8}
e. 9.6×10^{-3}; 1.5×10^{-2}
f. 21×10^{3}; 2.1×10^{4}

4.13 Three different reactions have the following rates:

Reaction *A*: 2.4×10^{-4} moles per liter per second

Reaction *B*: 3.6×10^{-4} moles per liter per second

Reaction *C*: 6.5×10^{-5} moles per liter per second

Which of these reactions is the fastest? the slowest?

4.4 Arrange the following numbers in increasing order, starting with the smallest:

2.0×10^{-5}, 4.6×10^{3}, 3.6×10^{-1}, 7.1×10^{-5}, 6.0×10^{-2}

4.14 The solubility products (K_{sp}) of several ionic compounds are listed below. Arrange these compounds in order of increasing K_{sp}.

Compound	K_{sp}
$Al(OH)_3$	5×10^{-33}
$Cr(OH)_3$	1×10^{-30}
$Fe(OH)_2$	1×10^{-13}
$Fe(OH)_3$	5×10^{-38}
$Mg(OH)_2$	1×10^{-11}

4.5 Carry out the indicated operations, citing answers in standard exponential notation:

a. $(6.20 \times 10^4) \times (1.50 \times 10^8)$

b. $(4.2 \times 10^{-3}) \times (1.50 \times 10^8)$

c. $(3.62 \times 10^4) \times (2.91 \times 10^{-7})$

d. $(8.16 \times 10^{-4}) \times (4.78 \times 10^{19})$

e. $(1.39 \times 10^7)/(1.10 \times 10^4)$

4.6 Carry out the indicated operations:

a. $(3.48 \times 10^3)/(6.72 \times 10^5)$

b. $(7.2 \times 10^{-3})/(3.6 \times 10^{-4})$

c. $(2.60 \times 10^4)/(7.70 \times 10^{-12})$

d. $\dfrac{(6.10 \times 10^4) \times (3.18 \times 10^{-4})}{(8.08 \times 10^7) \times (1.62 \times 10^{11})}$

4.7 Evaluate

a. $(2.16 \times 10^{-3})^2$

b. $(4.9 \times 10^4)^3$

c. $(6.0 \times 10^{-21})^2$

d. $(9.0 \times 10^6)^{1/2}$

e. $(8.4 \times 10^5)^{1/2}$

4.8 Find the value of

a. $(6.2 \times 10^{-2})^{1/2}$

b. $(1.62 \times 10^{-7})^{1/2}$

c. $(2.14 \times 10^{10})^{1/3}$

d. $(6.0 \times 10^4)^2 \times (3.0 \times 10^{-7})^{1/2}$

e. $(3.0 \times 10^7)^{1/5}$

4.15 The masses of atoms of four different elements are listed below. In each case, calculate the mass of 6.022×10^{23} atoms (one mole):

He	6.647×10^{-24} g
N	2.325×10^{-23} g
Sr	1.455×10^{-22} g

Which of these atoms is the heaviest? the lightest?

4.16 In water at 25° C, the concentrations of H^+ and OH^- are related by the equation

$$(\text{conc. } H^+) \times (\text{conc. } OH^-) = 1.0 \times 10^{-14}$$

Using this relation, complete the table below:

conc. H⁺	*conc. OH⁻*	$\dfrac{\textit{conc. } H^+}{\textit{conc. } OH^-}$
______	2.5×10^{-4}	______
3.6×10^{-8}	______	______
______	______	10^2

4.17 According to the Bohr theory of the H atom, the orbital radius is given by

$$r = \frac{n^2h^2}{4\pi^2me^2}$$

where r is the radius in centimeters, $h = 6.6 \times 10^{-27}$, $e = 4.8 \times 10^{-10}$, $m = 9.1 \times 10^{-28}$, $\pi = 3.14$, and n is the so-called quantum number which can have any integral positive value. Calculate r when $n = 1$; $n = 2$; $n = 3$.

4.18 When hypochlorous acid, HOCl, is dissolved in water, the following relation exists:

$$(\text{conc. } H^+)^2 = (3.2 \times 10^{-8}) \times (\text{conc. HOCl})$$

Complete the following table:

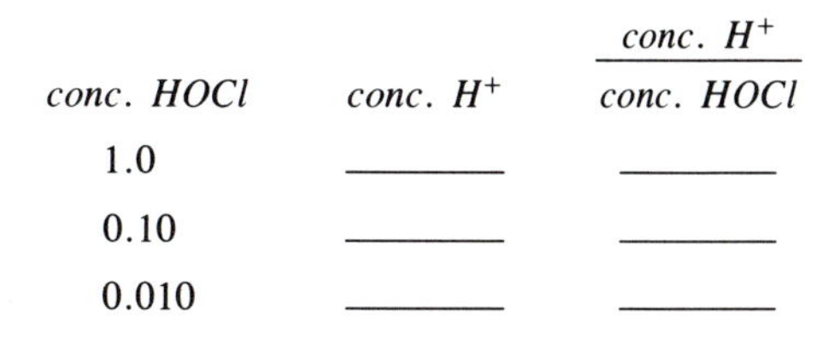

conc. HOCl	*conc. H⁺*	$\dfrac{\textit{conc. } H^+}{\textit{conc. HOCl}}$
1.0	______	______
0.10	______	______
0.010	______	______

4.9 Carry out the indicated operations:

a. $3.02 \times 10^4 + 1.69 \times 10^4$

b. $6.10 \times 10^4 + 1.0 \times 10^3$

c. $8.17 \times 10^5 - 1.20 \times 10^4$

d. $9.68 \times 10^4 + 7.01 \times 10^2$

4.10 Find the value of

a. $4.18 \times 10^{-2} + 1.29 \times 10^{-2}$

b. $5.9 \times 10^{-5} + 1.86 \times 10^{-4}$

c. $6.49 \times 10^{-10} - 1.23 \times 10^{-11}$

d. $6.02 \times 10^{23} - 1.0 \times 10^2$

4.19 Dalton's Law tells us that the total pressure of a mixture of hydrogen and helium is the sum of the partial pressures of the two gases. If the total pressure is 1.224×10^3 mm Hg and the partial pressure of helium is 9.80×10^2mm Hg, what is the partial pressure of hydrogen?

4.20 A certain reaction is carried out in a bomb calorimeter. Part of the heat evolved in the reaction is absorbed by the water, part by the bomb. If the water absorbs 12.02 kJ and the bomb absorbs 650 J, what is the total amount of heat evolved? (1 kJ = 10^3 J)

CHAPTER 5
Logarithms

Within the past decade, there has been a drastic change in the use of logarithms in general chemistry. Only a short time ago, a table of logarithms was an essential tool for chemical calculations. Accurate multiplications and divisions, valid to 4 or 5 digits, were commonly carried out using such a table. Students had to learn how to use a "log table" to obtain a logarithm or antilogarithm (number corresponding to a given logarithm). In the typical general chemistry course, probably more time was spent discussing "logs" and "antilogs" than any other concept in mathematics.

Today, thanks to the electronic calculator, the situation is quite different. Using your calculator, you can carry out a multiplication or division to 8 digits in a fraction of the time required with a table of logarithms. Indeed, "log tables" have, to all intents and purposes, gone the way of the buggy whip and the nickel candy bar. Logarithms and antilogarithms are readily obtained with an electronic calculator (Sections 5.1, 5.2).

These changes do not mean that you can safely skip this chapter. Many of the basic equations in chemistry include logarithmic terms (Table 5.2). To use these equations, you must know what logarithms are and how to work with them.

5.1 LOGARITHMS OF NUMBERS

The common logarithm of a number is defined quite simply. **It is the power to which 10 must be raised to give the number.** Thus,

$$\log 1000 = \log 10^3 = 3$$
$$\log 100 = \log 10^2 = 2$$
$$\log 10 = \log 10^1 = 1$$
$$\log 1 = \log 10^0 = 0$$
$$\log 0.1 = \log 10^{-1} = -1$$
$$\log 0.01 = \log 10^{-2} = -2$$
$$\log 0.001 = \log 10^{-3} = -3$$

In general, we can say that

$$\log 10^x = x \quad (5.1)$$

Notice from these examples that numbers greater than 1 (e.g., 10, 100, . . .) have positive logarithms (+1, +2, . . .). Numbers smaller than 1 (e.g., 0.1, 0.01, . . .) have negative logarithms (−1, −2, . . .). In general

$$\text{if } x > 1, \qquad \text{then } \log x > 0 \tag{5.2}$$

$$\text{if } x = 1, \qquad \text{then } \log x = 0 \tag{5.3}$$

$$\text{if } x < 1, \qquad \text{then } \log x < 0 \tag{5.4}$$

So far, all the numbers we have cited have been integral powers of 10. These numbers have logarithms which are integers, such as +2 or −1. It is possible to assign logarithms to all positive numbers.* For example, the number 2.00 has a logarithm; that is, 10 raised to *some* decimal power must be equal to 2.00. To find the logarithm of 2.00, all you need do is enter 2.00 on your calculator and press the "log" key:

Press	Display
2.00	**2.00**
log	**0.301 . .**

We conclude that

$$\log 2.00 = 0.301 \ldots$$

This means that $10^{0.301} = 2.00$. In other words, 0.301 is the power to which 10 must be raised to give 2.00. Similarly, you should be able to show, using your calculator, that

$$\log 3.00 = 0.477; \qquad 10^{0.477} = 3.00$$

$$\log 4.00 = 0.602; \qquad 10^{0.602} = 4.00$$

$$\log 5.00 = 0.699; \qquad 10^{0.699} = 5.00$$

Often you will need to find the logarithm of a number which is either too large or too small to be entered into your calculator. Ordinarily, such numbers will be in standard exponential notation (Chapter 4). Typical examples might be

$$2.00 \times 10^{15}; \qquad 2.00 \times 10^{-19}$$

To find the logarithms of these exponential numbers, you have two choices. One is to enter the number in exponential notation on your calculator, as described in Chapter 4. The logarithm can then be taken in the usual way. The other approach makes use of the general rule

$$\log (C \times 10^n) = n + \log C \tag{5.5}$$

*Negative numbers cannot be assigned logarithms. It is impossible to obtain a negative number by raising 10 to any power whatsoever.

where C is the coefficient of the exponential number and n is the exponent. Thus,

$$\log (2.00 \times 10^{15}) = 15 + \log 2.00$$

$$\log (2.00 \times 10^{-19}) = -19 + \log 2.00$$

Since the logarithm of 2.00 is 0.301, we have

$$\log (2.00 \times 10^{15}) = 15 + 0.301 = 15.301$$

$$\log (2.00 \times 10^{-19}) = -19 + 0.301 = -18.699$$

Example 5.1 Find the logarithms of

a. 7.15 b. 71.5 c. 7.15×10^{9} d. 7.15×10^{-9}

Solution

a. Using your calculator, you should find that log 7.15 = 0.854 . .
b. Again, using your calculator: log 71.5 = 1.854 . . Notice that log 71.5 is greater by 1 than log 7.15 (i.e., 1.854 vs. 0.854). This is precisely what you should expect from Equation 5.5, since $71.5 = 7.15 \times 10^{1}$.
c. Applying Equation 5.5,

$$\log (7.15 \times 10^{9}) = 9 + \log 7.15 = 9 + 0.854 = 9.854$$

d. Using Equation 5.5,

$$\log (7.15 \times 10^{-9}) = -9 + \log 7.15 = -9 + 0.854 = -8.146$$

Alternatively, you could use exponential notation on your calculator. The steps are as follows:

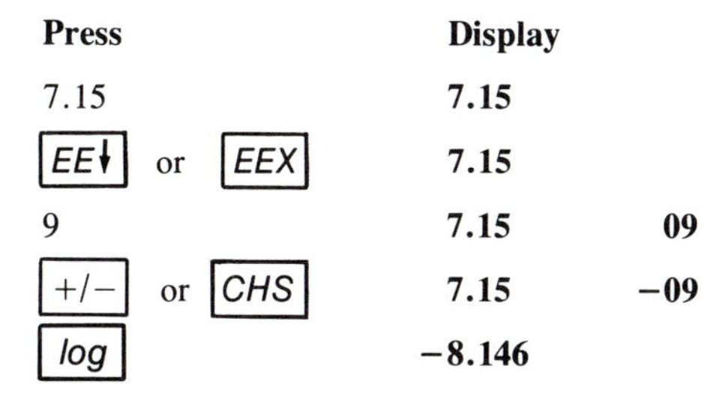

Press	**Display**	
7.15	**7.15**	
EE↓ or EEX	**7.15**	
9	**7.15**	**09**
+/− or CHS	**7.15**	**−09**
log	**−8.146**	

(The logarithm in part c could have been found the same way; try it.)

5.2 ANTILOGARITHMS

We can think of an antilogarithm as **the number corresponding to a given logarithm.** Thus,

$$\log 10^{3} = 3; \qquad \text{antilog } 3 = 10^{3} = 1000$$

$$\log 10^{2} = 2; \qquad \text{antilog } 2 = 10^{2} = 100$$

$$\log 10^{1} = 1; \qquad \text{antilog } 1 = 10^{1} = 10$$

$$\log 10^0 = 0; \qquad \text{antilog } 0 = 10^0 = 1$$

$$\log 10^{-1} = -1; \qquad \text{antilog } -1 = 10^{-1} = 0.1$$

$$\log 10^{-2} = -2; \qquad \text{antilog } -2 = 10^{-2} = 0.01$$

$$\log 10^{-3} = -3; \qquad \text{antilog } -3 = 10^{-3} = 0.001$$

In general,

$$\mathbf{\log 10^x = x; \qquad antilog\ x = 10^x} \tag{5.6}$$

From these examples, we see that it is possible to find antilogarithms for either positive or negative numbers. The antilog of a positive number (e.g., +2, +1, . . .) will be greater than 1 (e.g., 100, 10, . . .). The antilog of a negative number (e.g., −2, −1, . . .) must be less than 1 (e.g., 0.01, 0.1, . . .). In general,

$$\text{if } x > 0, \qquad \text{antilog } x > 1 \tag{5.7}$$

$$\text{if } x = 0, \qquad \text{antilog } x = 1 \tag{5.8}$$

$$\text{if } x < 0, \qquad \text{antilog } x < 1 \tag{5.9}$$

Equation 5.6 can be used directly to find the antilogarithm of a positive or negative *integer*. Thus,

$$\text{antilog } 62 = 10^{62}; \qquad \text{antilog } -51 = 10^{-51}$$

More often, you will need to find the antilogarithm of a decimal number such as 2.306 or −1.853. Here, you use your calculator. The procedure followed depends upon the type of calculator. The two most common methods use the basic relations given by Equation 5.6. One of these (followed by Hewlett-Packard and many other manufacturers) uses a 10^x key to find antilogarithms. The other approach (certain Texas Instruments calculators) uses the fact that the antilogarithm is the "inverse function" of the logarithm. Here, you press the "INV" and "log" keys in that order.

To illustrate how to find antilogs, suppose you want the antilogarithm of 2.306 (i.e., the number whose logarithm is 2.306). Depending on your brand of calculator, you follow one or the other of the sequences:

Press	Display	Press	Display
2.306	**2.306**	2.306	**2.306**
10^x	**202 . .**	INV	**2.306**
		log	**202 . .**

We conclude that $10^{2.306} = 202$, *or* the number whose logarithm is 2.306 is 202, *or* 202 is the antilogarithm of 2.306. All of these statements are true and indeed mean the same thing.

The procedure used to obtain antilogarithms of negative numbers involves one extra step. You have to use the "change sign" key, labeled +/− or CHS

To illustrate, suppose you want to find the antilogarithm of −1.853:

Press	Display	Press	Display
1.853	**1.853**	1.853	**1.853**
+/− or CHS	**−1.853**	+/−	**−1.853**
10^x	**0.0140 . .**	INV	**−1.853**
		log	**0.0140 . .**

The antilogarithm of −1.853 is 0.0140 (i.e., 1.40×10^{-2}).

Example 5.2 Find the numbers whose logarithms are

a. 0.8641 b. −0.8641 c. 116

Solution

a. Using your calculator, you should find that

$$\text{antilog } 0.8641 = 7.313\ .\ .$$

Note that, as predicted by Equation 5.7, the antilog of a positive number, 0.8641, is a number greater than one, 7.313.

b. Again using your calculator, with the "change sign" key this time,

$$\text{antilog of } -0.8641 = 0.1367\ .\ .$$

This illustrates Equation 5.9; the antilog of a negative number, −0.8641, is a number less than one, 0.1367.

c. If you attempt to use your calculator here, you will get an "error" signal. Calculators will not handle antilogarithms of numbers greater than 100 or smaller than −100. However, the answer is readily obtained from Equation 5.6:

$$\text{antilog } 116 = 10^{116}$$

5.3 OPERATIONS INVOLVING LOGARITHMS

Since logarithms are exponents, the rules derived in Chapter 4 for working with exponents can be extended to logarithms. The results are summarized in Table 5.1.

Table 5.1 MATHEMATICAL OPERATIONS INVOLVING EXPONENTS AND LOGARITHMS

	EXPONENTS	LOGARITHMS
Multiplication	$10^a \times 10^b = 10^{(a+b)}$	$\log (xy) = \log x + \log y$
Division	$10^a/10^b = 10^{(a-b)}$	$\log (x/y) = \log x - \log y$
Raising to a power	$(10^a)^n = 10^{an}$	$\log x^n = n \log x$
Extracting a root	$(10^a)^{1/n}$ $10^{a/n}$	$\log x^{1/n} = \frac{1}{n} \log x$

The rules given in Table 5.1 are easily checked on your calculator.

MULTIPLICATION. According to Table 5.1,

$$\log (xy) = \log x + \log y$$

If we let $x = 3.00$ and $y = 2.00$, then $xy = 6.00$.

$$\log 6.00 = \log 3.00 + \log 2.00$$

From your calculator:

$$\log 6.00 = 0.778; \qquad \log 3.00 = 0.477; \qquad \log 2.00 = 0.301$$

$$0.778 = 0.477 + 0.301$$

DIVISION. According to Table 5.1,

$$\log (x/y) = \log x - \log y$$

Again, if we let $x = 3.00$ and $y = 2.00$, then $x/y = 1.50$.

$$\log 1.50 = \log 3.00 - \log 2.00$$

From your calculator:

$$\log 1.50 = 0.176; \qquad \log 3.00 = 0.477; \qquad \log 2.00 = 0.301$$

$$0.176 = 0.477 - 0.301$$

RAISING TO A POWER

$$\log x^n = n \log x$$

If we let $x = 3.00$ and $n = 2$, then $x^n = (3.00)^2 = 9.00$.

$$\log 9.00 = 2 \times \log 3.00$$

But $\log 3.00 = 0.477; \qquad \log 9.00 = 0.954$

$$2 \times 0.477 = 0.954$$

EXTRACTING A ROOT

$$\log x^{1/n} = \frac{1}{n} \log x$$

If we let $x = 4.00$ and $n = 2$, then $x^{1/n} = (4.00)^{1/2} = 2.00$.

$$\log 2.00 = \tfrac{1}{2} \log 4.00$$

But $\log 2.00 = 0.301; \qquad \log 4.00 = 0.602$

$$0.301 = \tfrac{1}{2} (0.602)$$

In chemistry, operations such as those listed in Table 5.1 are most often carried out with physical quantities (pressure, concentration, and so on) rather than numbers. Example 5.3 illustrates this application of the laws governing the use of logarithms, using x, y, and z as symbols for quantities.

Example 5.3 Using the rules given in Table 5.1,

a. Express $\log \frac{xy^{1/2}}{z^2}$ in terms of the logarithms of x, y, and z.

b. Express the quantity $2 \log x + \log y - \frac{1}{2} \log z$ as the logarithm of a single quantity (a function of x, y, and z).

Solution

a. Following the rules for multiplication and division,

$$\log \frac{xy^{1/2}}{z^2} = \log x + \log y^{1/2} - \log z^2$$

Applying the rules for raising to a power and extracting roots,

$$= \log x + \tfrac{1}{2} \log y - 2 \log z$$

b. Here, we reverse the operations in (a). We start by applying the rules for raising to a power and extracting roots:

$$2 \log x = \log x^2; \qquad \tfrac{1}{2} \log z = \log z^{1/2}$$

so

$$2 \log x + \log y - \tfrac{1}{2} \log z = \log x^2 + \log y - \log z^{1/2}$$

Now we apply the rules for multiplication and division:

$$= \log \frac{x^2y}{z^{1/2}}$$

5.4 NATURAL LOGARITHMS

Up to this point, we have been discussing "common logarithms," i.e., logarithms to the base 10. For calculation purposes, common logarithms are the simplest to work with, since our number system is based on multiples of 10. However, many of the equations we use in general chemistry are expressed more simply in terms of a different type of logarithm, taken to the base ***e***, where:

$$e = 2.718 \ldots$$

Logarithms to the base e are referred to as natural logarithms. To distinguish natural from common logarithms, the abbreviation **ln** is used:

$$\log_e x \equiv \ln x; \qquad \log_{10} x \equiv \log x$$

Natural logarithms are readily found on your calculator, using the [*ln x*] key.

You should find that

$$\ln 10.00 = 2.303 \qquad \log 10.00 = 1.000$$

$$\ln 5.00 = 1.609 \qquad \log 5.00 = 0.699$$

$$\ln 2.00 = 0.693 \qquad \log 2.00 = 0.301$$

$$\ln 1.00 = 0.000 \qquad \log 1.00 = 0.000$$

$$\ln 0.500 = -0.693 \qquad \log 0.500 = -0.301$$

$$\ln 0.200 = -1.609 \qquad \log 0.200 = -0.699$$

$$\ln 0.100 = -2.303 \qquad \log 0.100 = -1.000$$

From these examples and others we can show that

1. $$\ln x = \mathbf{2.303} \log x \qquad (5.10)$$

(Note, for example, that ln 1.00/log 1.00 = 2.303/1.000 = 2.303; ln 5.00/log 5.00 = 1.609/0.699 = 2.303). This equation is often useful for converting from base 10 to natural logarithms or vice versa.

2. Equations 5.2–5.4 apply to natural as well as base 10 logarithms:

$$\text{if } x > 1, \quad \ln x > 0 \qquad (\ln 2.00 = 0.693)$$

$$\text{if } x = 1, \quad \ln x = 0$$

$$\text{if } x < 1, \quad \ln x < 0 \qquad (\ln 0.200 = -1.609)$$

3. The rules in Table 5.1 apply to natural as well as base 10 logarithms. Take, for example, the rule that $\log (xy) = \log x + \log y$. Noting that $1 = 5 \times 0.2$, we have

$$\ln 1 = \ln 5 + \ln 0.2; \qquad 0.000 = 1.609 - 1.609$$

Example 5.4 Find the value of

a. ln 2.7183 b. $\ln(2.02 \times 10^{-16})$ c. $\ln(x^{1/3}y^2)$, in terms of $\ln x$ and $\ln y$

Solution

a. On your calculator you should find that

$$\ln 2.7183 = 1.0000$$

Recalling that $e = 2.718 \ldots$, it is reasonable that $\ln e = 1$, just as $\log 10 = 1$.

b. The natural logarithm of 2.02×10^{-16} is *not* equal to $-16 + \ln 2.02$ (why?). Probably the simplest way to carry out this operation is to use exponential notation on your calculator. In this way, you should find that

$$\ln (2.02 \times 10^{-16}) = -36.138$$

c. Proceeding as in Example 5.3a,

$$\ln(x^{1/3}y^2) = \ln x^{1/3} + \ln y^2 = \tfrac{1}{3} \ln x + 2 \ln y$$

From time to time you may need to evaluate expressions in which the base of natural logarithms, e, is raised to a power. This is easily accomplished if you have an " e^x " key on your calculator. With certain Texas Instruments calculators, the operation is not quite so obvious. Here, you use the fact that e^x is the "inverse function" of $\ln x$. Hence, you press the " INV " and " ln x " keys in that order. Either way, you should find that

$$
\begin{aligned}
e^2 &= 7.389\ldots \\
e^1 &= 2.718\ldots \\
e^{1/2} &= 1.649\ldots \\
e^0 &= 1.000\ldots \\
e^{-1/2} &= 0.6065\ldots \\
e^{-1} &= 0.3679\ldots \\
e^{-2} &= 0.1353\ldots
\end{aligned}
$$

The base of natural logarithms, e, can be expressed as the sum of an infinite series:

$$
\begin{aligned}
e &= 1 + \frac{1}{1} + \frac{1}{1 \times 2} + \frac{1}{1 \times 2 \times 3} + \frac{1}{1 \times 2 \times 3 \times 4} + \frac{1}{1 \times 2 \times 3 \times 4 \times 5} + \cdots \\
&= 1.000 + 1.000 + 0.500 + 0.167 + 0.042 + 0.008 + 0.001 + \cdots \\
&= 2.718\ldots
\end{aligned}
$$

Natural logarithms can also be evaluated by using infinite series. One such series is

$$\ln x = 2\left[\frac{x-1}{x+1} + \frac{1}{3}\left(\frac{x-1}{x+1}\right)^3 + \frac{1}{5}\left(\frac{x-1}{x+1}\right)^5 + \frac{1}{7}\left(\frac{x-1}{x+1}\right)^7 + \cdots\right]$$

To evaluate the natural logarithm of 2, we could write, with $x = 2$,

$$
\begin{aligned}
\ln 2 &= 2\left[\frac{1}{3} + \frac{1}{3}\left(\frac{1}{3}\right)^3 + \frac{1}{5}\left(\frac{1}{3}\right)^5 + \frac{1}{7}\left(\frac{1}{3}\right)^7 + \cdots\right] \\
&= 2[0.33333\ldots + 0.01235\ldots + 0.00082\ldots + 0.00007\ldots] \\
&= 0.69314\ldots
\end{aligned}
$$

In your calculator, a routine is activated when you press the " ln x " key. This carries out a summation similar to the one we have indicated to obtain the natural logarithm to 8 to 10 digits. All this happens in less than one second, which is rather remarkable when you think about it. A calculator obtains base 10 logarithms by dividing by 2.302585093, which is the natural logarithm of 10.

$$\log 2 = \frac{0.69314\ldots}{2.302585093} = 0.30103\ldots$$

5.5 APPLICATIONS IN CHEMISTRY

The most common use of logarithms in general chemistry is in the treatment of acids and bases. One way of describing the acidity or basicity of a solution is to cite a quantity known as pH, defined as follows:

$$\text{pH} = -\log(\text{conc. H}^+) \qquad (5.11)$$

where "conc. H^+" stands for the concentration of hydrogen ions, H^+, the cation responsible for acidic properties. This equation is often used to calculate the pH of a solution, knowing the concentration of H^+. It can also be used to carry out the reverse calculation, going from pH to conc. H^+. Both calculations are readily carried out on your calculator. The first requires that you find a logarithm, the second an antilogarithm. Be sure to remember the minus sign in Equation 5.11!

Example 5.5 Find

a. the pH of a solution in which conc. $H^+ = 2.0 \times 10^{-3}$
b. the pH of a solution in which conc. $H^+ = 2.0 \times 10^{-14}$
c. conc. H^+ in a solution in which pH $= 2.81$
d. conc. H^+ in a solution in which pH $= -1.00$

Solution

a. There are three somewhat different ways of working this problem:

— enter 0.0020 on your calculator and press the [*log*] key. You should find that log 0.0020 $= -2.70$. Hence, pH $= -\log(\text{conc. H}^+) = -(-2.70) = 2.70$.

— enter 2.0×10^{-3} on your calculator, press the [*log*] key and continue as above.

— apply Equation 5.5:

$$\log(2.0 \times 10^{-3}) = -3 + \log 2 = -3 + 0.30 = -2.70$$

$$\text{pH} = -(-2.70) = 2.70$$

b. Here you can either use your calculator in exponential notation or apply Equation 5.5:

$$\log(2.0 \times 10^{-14}) = -14 + \log 2 = -14 + 0.30 = -13.70$$

$$\text{pH} = -\log(2.0 \times 10^{-14}) = 13.70$$

c. If pH $= 2.81$, then log (conc. H^+) $= -2.81$. To find conc. H^+, you must find the antilogarithm of -2.81. Using your calculator,

$$\text{antilog of } -2.81 = 0.0015 = 1.5 \times 10^{-3} = \text{conc. H}^+$$

d. pH $= -1.00$; log (conc. H^+) $= 1.00$; conc. $H^+ = 10.0$

Many of the basic principles of general chemistry are expressed in equations which include logarithmic terms. Some of these equations are listed in Table 5.2. Note that the number 2.303 appears in each equation. This is the factor that converts natural to base 10 logarithms. The equations would be somewhat simpler if natural logarithms were used. For example, the first equation in Table 5.2 would read:

$$\Delta G^\circ = -RT \ln K$$

Table 5.2 SOME EQUATIONS OF GENERAL CHEMISTRY INVOLVING LOGARITHMIC TERMS

EQUATION	*MEANING OF SYMBOLS*
1. $\Delta G^\circ = -2.303\ RT \log K$	ΔG° = std. free energy change (joules) R = gas constant = 8.314 J/(mol · K) T = temperature in degrees Kelvin K = equilibrium constant
2. $\log \frac{X_0}{X} = \frac{kt}{2.303}$ (1st order rate law)	X_0 = original concentration of reactant X = concentration at time t k = 1st order rate constant
3. $\log \frac{P_2}{P_1} = \frac{\Delta H(T_2 - T_1)}{2.303\ RT_2T_1}$	P_2, P_1 = vapor pressure of liquid at absolute temperatures T_2 and T_1 ΔH = heat of vaporization of liquid in joules per mole
4. $\log \frac{K_2}{K_1} = \frac{\Delta H(T_2 - T_1)}{2.303\ RT_2T_1}$	K_2, K_1 = equilibrium constants at temperatures T_2 and T_1 ΔH = enthalpy change in reaction in joules
5. $\log \frac{k_2}{k_1} = \frac{\Delta E_a(T_2 - T_1)}{2.303RT_2T_1}$	k_2, k_1 = rate constants at temperatures T_2 and T_1 ΔE_a = activation energy in joules

In the "word problems" starting on p. 64, we will illustrate the use of several of the equations listed in Table 5.2. As indicated in Example 5.6, you should start by writing down the appropriate equation. If necessary, solve this equation for the quantity required. Then substitute numbers and solve for a numerical answer, using your calculator, in one continuous operation, if possible.

Example 5.6 Using Equation 1 in Table 5.2, calculate

a. ΔG° (in joules) for a reaction in which the equilibrium constant is 0.1000 at T = 300.0 K.
b. $\log K$ for a reaction in which $\Delta G^\circ = +2.120 \times 10^4$ J at T = 1000 K.
c. K for the reaction in (b).

Solution

a. $G^\circ = -2.303\ RT \log K$
$= -2.303(8.314)(300.0) \log (0.1000) = +5.744 \times 10^3$ J
b. Solving Equation 1 for $\log K$,

$$\log K = \frac{-\Delta G^\circ}{2.303\ \mathrm{RT}} = \frac{-21{,}200}{(2.303)(8.314)(1000)} = -1.107$$

c. We need to find the number whose logarithm is -1.107:
antilog of $-1.107 = 0.0782$
hence $K = 0.0782 = 7.82 \times 10^{-2}$

PROBLEMS

5.1 Find the logarithms of the following numbers:

a. 8.16 *b.* 1.652 *c.* 0.004918
d. 1.0×10^{-6} *e.* 4.5×10^{-10}

5.2 Find the numbers whose logarithms are

a. 0.8831 *b.* 1.743 *c.* −3.161
d. −6.918 *e.* −126

5.3 Calculate the value of y in each case:

a. $y = -6.14 \times 100 \times \log 0.0360$

b. $54.0 = \frac{116.2}{3.98} \log y$

c. $-1.614 = (300y) \log (2.0 \times 10^3)$

5.4 Calculate y in each case:

a. $\log 12 = \frac{0.019y}{2.30}$

b. $\log y = \frac{(1.2 \times 10^{-3})(3600)}{2.30}$

c. $\log \frac{3.0}{y} = 0.16$

5.5 Find the value of log *y* if

a. $y = 1.6 \times 10^{-4}$
b. $y^2 = 1.6 \times 10^{-4}$
c. $y^{1/2} = 1.6 \times 10^{-4}$

5.6 Express in terms of log *x* and log *y*:

a. $\log x/y$ *b.* $\log xy$ *c.* $\log xy^2$

5.7 Given that

$$x = y^2/z^{1/2}$$

obtain a relation between log *x*, log *y*, and log *z*.

5.11 Calculate the pH of solutions which have the following concentrations of H^+:

a. 1.0×10^{-6} *b.* 2.61×10^{-2}
c. 3.0×10^{-9} *d.* 6.0

5.12 Calculate the concentration of H^+ in solutions with the following pH:

a. 4.0 *b.* 12.60 *c.* 3.14 *d.* −1.0

5.13 Using Equation 1, Table 5.2, calculate

a. $\Delta G°$ at 300 *K* for a reaction for which the equilibrium constant *K* is 0.020.

b. the equilibrium constant for a reaction at 400 *K* where $\Delta G° = +1.08 \times 10^4$J.

5.14 Using Equation 2, Table 5.2, calculate

a. the time required for the concentration of reactant to drop from 1.0 to 0.10 if $k = 0.045$/min.

b. the concentration of reactant after 50 min, starting with a concentration of 2.00 and taking $k = 2.0 \times 10^{-3}$/min.

5.15 In a water solution saturated with H_2S;

$$(\text{conc. } H^+)^2 \times (\text{conc. } S^{2-}) = 1 \times 10^{-21}$$

What is the concentration of S^{2-} in a solution of this type which has a pH of 4.0?

5.16 In any water solution at 25°C;

$$(\text{conc. } H^+) \times (\text{conc. } OH^-) = 1.0 \times 10^{-14}$$

Show, using this equation, that

$$\text{pH} + \text{pOH} = 14.00$$

where pOH = −log (conc. OH^-).

5.17 For a water solution of acetic acid, the following relation holds:

$$K_a = \frac{(\text{conc. } H^+)^2}{(\text{conc. HAc})}$$

where K_a is the ionization constant of acetic acid, HAc. Starting with this equation, obtain a relation between log K_a, pH, and log (conc. HAc).

5.8 Consider the equation

$$y = 12.4 - 2.0 \log x^{8/5}$$

a. What is y when $x = 3.0$?

b. What is x when $y = 1.0$?

5.9 Find the natural logarithms of

a. 6.023 *b.* $(2.718)^2$

c. 6.18×10^{-5}

5.10 Find the value of

a. $e^{1.00}$ *b.* $e^{-2.50}$ *c.* $e^{12.92}$

5.18 In a certain electrical cell, the voltage, E, is given by the expression

$$E = +0.75\ V + \frac{0.0591}{5} \log (\text{conc. } H^+)^8$$

a. Write an equation relating E to pH.

b. Calculate E when pH $= 3.0$.

c. Calculate pH when $E = 0.50\ V$.

5.19 Equation 1 in Table 5.2 can be written in the form

$$\Delta G^\circ = -RT \ln K$$

Working with natural logarithms, calculate

a. ΔG° when $T = 250$, $K = 16$.

b. K when $\Delta G^\circ = -1.00 \times 10^4$J, $T = 100$.

5.20 The fraction, f, of molecules with energy equal to or greater than the activation energy, E_a, is given by the equation

$$f = e^{-E_a/RT}$$

What is f when $E_a = 1.00 \times 10^4$J, $T = 298$ K, $R = 8.31$ J/K?

CHAPTER 6

Significant Figures

The numbers that we work with in general chemistry can be divided into two broad categories. Some numbers are, by their nature, **exact.** When we refer to *two* crucibles or *five* beakers, we indicate the exact number of such items. Similarly, in the relations

$$1 \text{ ft} = 12 \text{ in.}; \qquad 1\ \ell = 1000 \text{ cm}^3$$

the numbers 12 and 1000 are exact. There are precisely *twelve* inches in one foot and *one thousand* cubic centimeters in one liter. Other numbers are **inexact.** When we refer to a "100-cm^3" beaker, we do not imply that it has a volume of *exactly* 100 cubic centimeters. If we fill the beaker with water, we may find that it holds as little as 90 cm^3 or as much as 110 cm^3.

Numbers which arise from measurements are always inexact. They always have associated with them an error:

$$\text{error} = \text{measured value} - \text{true value} \qquad (6.1)$$

or "uncertainty." Sometimes, the uncertainty is indicated by using the $\pm$ symbol. Suppose, for example, you weigh out a sample of sodium chloride on a crude balance and establish its mass to within 0.01 g. You might report that it weighed

$$2.65 \pm 0.01 \text{ g}$$

The understanding here is that there is an uncertainty or likely error of $\pm$ 0.01 g.

To find the mass of the sample more precisely, you could use a balance with a sensitivity of 0.001 g. This way, you could reduce the uncertainty to $\pm$ 0.001 g. You might find now that the sodium chloride weighed

$$2.652 \pm 0.001 \text{ g}$$

Analytical balances capable of weighing to $\pm$0.0001 g are available. Using such a balance, you might report the mass of the sodium chloride sample to be

$$2.6518 \pm 0.0001 \text{ g}$$

For a person who is trying to estimate the validity of an experiment or repeat it in another laboratory, it is important to specify the uncertainty of a measurement. One way to do this is to use the notation just shown. Often, though, we omit the ±0.01, ±0.001, and so on, and simply report

2.65 g; 2.652 g; 2.6518 g

Here, it is understood that there is an *uncertainty of one unit in the last digit*. When we say that a sample of sodium chloride weighs "2.65 g," we mean that its mass is between 2.64 and 2.66 g. Indeed, it must be closer to 2.65 than to 2.64 or 2.66 g.

The precision of a measurement can also be described in terms of the number of **significant figures.** We say that in "2.65 g" there are three significant figures; each of the three digits is experimentally meaningful. The masses "2.652 g" and "2.6518 g" are quoted to 4 and 5 significant figures. The greater the number of significant figures, the greater the precision and the smaller the uncertainty. This applies not only to individual measurements but also to quantities calculated from such measurements. In this chapter, we will consider the rules that govern the use of significant figures in measurements and in calculations.

6.1 COUNTING SIGNIFICANT FIGURES

Frequently, we are faced with the problem of deciding how many significant figures there are in a number reported in a textbook, laboratory manual, or other source. In many cases, there is no ambiguity. When we find the atomic mass of calcium listed as 40.08, we trust that it is known to four significant figures.

When either the first or the last digit in a quantity is zero, the number of significant figures may not be obvious. Here, common sense is a useful guide. When we find the atomic mass of krypton listed as 83.80, it should be clear that it is known to four significant figures. If the zero were not significant, there would be no reason for including it. Writing "83.80" implies that the true value of the atomic mass of krypton lies between 83.79 and 83.81.

In another case, suppose we are told that the volume of a gas is 0.02461 ℓ. Is the zero to the right of the decimal point significant? A moment's reflection should convince you that it is not. In this case, *the zero is used simply to fix the position of the decimal point*. To make this conclusion more obvious, we could express the volume in cubic centimeters rather than liters. Since 1 ℓ = 1000 cm^3, the volume would now be 24.61 cm^3, with four significant figures clearly indicated. We cannot change the precision of a measurement by changing its units. Hence, there must be four significant figures in the quantity 0.02461 ℓ.

Suppose we are asked to weigh out 500 g of sodium chloride for an experiment. To how many significant figures should we make this measurement: one? two? three? Unfortunately, we cannot answer this question without reading the mind of the person who wrote the directions for the experiment. Perhaps we are supposed to weigh out roughly a quantity between 400 and 600 g, i.e., 500 ± 100 g. If so, there is one significant figure. On the other hand, we may be expected to weigh to ±10 g, giving two significant figures. Then again, maybe we are supposed to weigh to the nearest gram. In that case, the number

"500" has three significant figures. About all we can do in cases like this is to wish that the mass had been expressed in standard exponential notation as either

$$5 \times 10^2 \text{ g} \quad \text{(1 significant figure)}$$

or

$$5.0 \times 10^2 \text{ g} \quad \text{(2 significant figures)}$$

or

$$5.00 \times 10^2 \text{ g} \quad \text{(3 significant figures)}$$

Had this been done, there would have been no ambiguity.

If you think about it, the argument we just went through leads to a general rule for determining the number of significant figures in a cited quantity. We need only express that quantity in standard exponential notation (Chapter 4), i.e., in the form

$$C \times 10^n$$

where C is a number between 1 and 10 and n is the appropriate power of 10. The number of digits in the coefficient, C, will equal the number of significant figures in the quantity. Thus we have

$$0.02461\ \ell = 2.461 \times 10^{-2}\ \ell; \quad \text{4 significant figures}$$

$$0.0045630 \text{ kg} = 4.5630 \times 10^{-3} \text{ kg}; \quad \text{5 significant figures}$$

Example 6.1 How many significant figures are there in

a. 1.204×10^{-2} g b. 3.160×10^7 nm c. 0.00281 g d. 810 g

Solution

a. All digits are significant; 4
b. All digits are significant; 4
c. Three. The zeros serve only to fix the position of the decimal point, as you can see by converting to standard exponential notation: 0.00281 g = 2.81×10^{-3} g.
d. Perhaps 2, in which case the mass should have been written as 8.1×10^2 g. Could also be 3 (8.10×10^2 g). This illustrates the advantage of expressing experimental measurements in standard exponential notation.

Occasionally, you may be asked a question such as "How many grams of carbon dioxide can be formed from one gram of carbon?" Here, where the number one is written out, you should assume that it is exact. In effect, you are being asked how much carbon dioxide can be produced from *exactly* one gram of carbon, i.e., 1.000 000 000 . . g of carbon. In that sense, the quantities indicated by "one gram" or "ten kilograms" are similar to the exact numbers "1" and "12" in the relation

$$1 \text{ ft} = 12 \text{ in.}$$

6.2 MULTIPLICATION AND DIVISION

Most of the quantities that we measure in the laboratory are not end results in themselves. Instead, they are used to calculate other quantities. We may, for example, multiply the length of a piece of tin foil by its width to determine its area. In another case, we may divide the mass of a sample by its volume to determine its density.

When we multiply or divide two experimental quantities, both of which are inexact, the product or quotient is inexact. The question arises as to how great an error arises from these operations. To answer this question, it will be helpful to work through a specific example. Let us suppose that the mass and volume of a sample are, in that order, 5.80 ±0.01 g and 2.6 ±0.1 cm^3. The density, found by dividing mass by volume, should be about (5.80/2.6) g/cm^3. However, taking into account the uncertainties in mass and volume, the density might be as large as (5.81/2.5) g/cm^3 or as small as (5.79/2.7) g/cm^3. Carrying out these divisions, we obtain

$$\frac{5.81}{2.5} = 2.32\ldots \qquad \frac{5.80}{2.6} = 2.23\ldots \qquad \frac{5.79}{2.7} = 2.14\ldots$$

Comparing these three quotients, we conclude that the density is 2.2 ±0.1 g/cm^3. In other words, the density, like the volume, is known to 2 significant figures.

This example illustrates a general rule for the multiplication or division of inexact numbers. In general, we should **retain in the product or quotient the number of significant figures in the least precise of the numbers.** Applying this rule, we deduce that

$(6.10 \times 10^3)(2.08 \times 10^{-4})$	has 3 significant figures in answer
5.92×3.0	has 2 significant figures in answer
$8.2/3.194$	has 2 significant figures in answer

If you carry out these operations on your calculator, you will find that there are many extra digits in the display. For example, direct multiplication of 6.10×10^3 by 2.08×10^{-4} gives

$$12.688 \times 10^{-1} \qquad \text{or } 1.2688$$

This number should be rounded off to three significant figures to give 1.27 as the answer. Again, direct division of 8.2 by 3.194 would give

$$\frac{8.2}{3.194} = 2.5673137\ldots$$

It would be absurd to report this eight digit number as your answer. You should round it off to two significant figures, giving 2.6. In general, whenever you multiply or divide with your calculator, you are very likely to obtain extra, meaningless digits. Make sure you round off to the correct number of significant figures.

Example 6.2 Carry out the following operations involving inexact numbers, retaining the correct number of significant figures:

a. 6.19×2.8
b. $3.18/1.702$
c. $(4.10 \times 3.02 \times 10^9)/1.5$

Solution

a. We should retain only two significant figures:

$$6.19 \times 2.8 = 17$$

b. $3.18/1.702 = 1.87$ (3 significant figures)
c. $(4.10 \times 3.02 \times 10^9)/1.5 = 8.3 \times 10^9$ (2 significant figures)

Frequently, in carrying out multiplications or divisions, we use conversion factors which arise from relations that are defined exactly (1 kg = 1000 g) or can be expressed to any desired number of digits (1 lb = 453.6 . . g). The use of such conversion factors never affects the number of significant figures in an answer. The precision of the product or quotient depends only upon those of the experimental quantities that enter into the calculation (Example 6.3).

Example 6.3 Express the mass of a steak weighing 0.75 lb in grams and kilograms.

Solution

To find the mass in grams, we use the conversion factor 453.6 g/1 lb:

$$0.75 \text{ lb} \times \frac{453.6 \text{ g}}{1 \text{ lb}} = 3.4 \times 10^2 \text{ g} \qquad \text{(2 significant figures)}$$

To convert to kilograms:

$$3.4 \times 10^2 \text{ g} \times \frac{1 \text{ kg}}{1000 \text{ g}} = 3.4 \times 10^{-1} \text{ kg} \qquad \text{(2 significant figures)}$$

In both cases, it is the number of significant figures in the measured mass (0.75 lb) that determines the precision of the answer.

The rules of significant figures can be used as a guide in designing laboratory experiments. If two measurements are to be combined by multiplying or dividing, the precision of the result will be governed by that of the least precise measurement. Suppose we determine the density of a liquid using a triple beam balance (± 0.01 g) and a graduated cylinder (± 0.1 cm^3). The precision of the density will be limited by the large uncertainty of the volume measurement. If we need to know the density more exactly, we must use a more precise device for measuring volume, such as a pipet or buret (± 0.01 cm^3). We could not improve the precision by using an analytical balance weighing to ± 0.0001 g and then measure volume to ± 0.1 cm^3:

One precaution is in order concerning the rule we have given for significant figures in a product or quotient. It is an approximation which can sometimes lead to absurd situations. Let us suppose that two students, asked to determine the density of a liquid, obtain the following results:

	Mass	**Volume**	**Density**
Student 1	10.20 g	10.1 cm^3	1.01 g/cm^3
Student 2	10.10 g	9.9 cm^3	1.01 g/cm^3

Should the second student drop the last digit in his calculated density, reporting it as 1.0 g/cm^3, since there are only 2 significant figures in the volume? Common sense tells us that he should not. Like the other student, he has measured the volume to about ±1 percent. It is quite reasonable for him to report a density to ±1 percent. Thus he should report 1.01 g/cm^3, rather than 1.0 g/cm^3, which would imply an uncertainty of ±10 percent.

Situations as extreme as this do not arise too frequently, They do, however, point out the limitations of arbitrary rules such as those we have given for significant figures. There is an alternative way to estimate the uncertainty, or error, in a product or quotient. This involves the percent error in a measured quantity:

$$\text{Percent error} = \frac{\text{Error}}{\text{Measured value}} \times 100 \qquad (6.2)$$

*The percent error in a product or quotient is approximately equal to the percent error in the least precise number.** Applying this principle to the example discussed above (Student 2),

$$\text{mass} \quad 10.10\text{ g} \pm 0.01\text{ g}; \qquad \%\text{ error} = \frac{0.01}{10.10} \times 100 = 0.1$$

$$\text{volume} \quad 9.9\text{ cm}^3 \pm 0.1\text{ cm}^3; \qquad \%\text{ error} = \frac{0.1}{9.9} \times 100 = 1$$

$$\%\text{ error in density} \approx \%\text{ error in volume} = 1$$

$$\text{error in density} = \frac{1}{100} \times 1.01\text{ g/cm}^3 = 0.01\text{ g/cm}^3$$

$$\text{density} = 1.01\text{ g/cm}^3 \pm 0.01\text{ g/cm}^3$$

We see that the student is indeed justified in reporting the density to 3 significant figures.

6.3 ADDITION AND SUBTRACTION

When we add or subtract inexact numbers, we apply the following general principle: **A sum or difference cannot have an absolute precision greater than that of the least precise number involved.** To illustrate this point, suppose we add 1.32 g of sodium chloride and 0.006 g of potassium chloride to 28 g of water. How should we express the mass of the resulting solution? The implied precisions of these masses are 0.01 g, 0.001 g, and 1 g in that order.

*More exactly, $E^2 = E_1^2 + E_2^2$ where E_1 and E_2 are the percent errors in the two numbers and E is the percent error in their product or quotient. If one of the errors, let us say E_1, is much larger than the other, E_2, then: $E^2 \approx E_1^2$; $E \approx E_1$, which is the rule quoted above.

sodium chloride	1.32	±0.01 g
potassium chloride	0.006	±0.001 g
water	28	±1 g
total mass	29	±1 g

The sum of the masses cannot be more precise than that of the water (±1 g). We should write the total mass as 29 g, rather than 29.3 g, 29.326 g, or some other number of grams.

In another case, suppose we weigh out a sample of sodium chloride by difference. We start with a mass of 32.241 g and pour out enough sodium chloride to reduce the mass to 32.13 g. The mass of the sample is

$$\begin{array}{r} 32.241 \text{ g} \\ -32.13\ \ \text{ g} \\ \hline 0.11\ \ \text{ g} \end{array}$$

The mass is expressed only to the nearest hundredth of a gram, because that is the uncertainty in the least precise weighing (the last one).

You will note that in adding or subtracting inexact quantities, the principles are quite different from those governing multiplication and division. In particular, the number of significant figures in a sum or difference is *not* governed by the quantity having the fewest significant figures. Recall that in our first illustration the mass of potassium chloride was known to only one significant figure (0.006 g). Yet the total mass could be expressed to 2 significant figures (29 g). Again, in the second illustration, the mass of sodium chloride found by difference could be quoted to only 2 significant figures (0.11 g). This was true despite the fact that the final and initial weighings were made to 4 and 5 significant figures in that order. Cases such as these are common. We often "gain" significant figures in addition and "lose" them in subtraction. Remember, it is the *absolute* uncertainty (±1 g, ±0.01 g, etc.) which governs the number of significant figures in addition or subtraction.

Example 6.4 Carry out the following operations, giving answers to the correct number of significant figures:

a. 6.82 g + 2.111 g + 1268 g
b. 213 g − 0.01 g
c. 22.611 g − 22.534 g

Solution

a. Since "1268 g" has the greatest uncertainty, ±1 g, we write

$$7 \text{ g} + 2 \text{ g} + 1268 \text{ g} = 1277 \text{ g}$$

b. The mass 213 g is known only to ± 1 g, so subtracting 0.01 g from it has no effect; 213 g − 0 g = 213 g.
c. 22.611 g − 22.534 g = 0.077 g. Notice that although the two masses involved are both known to 5 significant figures, their difference is known only to 2. This general effect is one you should keep in mind in the laboratory. If you weigh out very small samples, say 0.1 g or less, your percent error is likely to be quite large. This is true despite the fact that the weighings themselves are very precise.

6.4 ROUNDING OFF NUMBERS

In calculations from experimental data, it is often necessary to drop one or more digits to obtain an answer with the appropriate number of significant figures. The rules which are followed in rounding off are:

1. If the first digit dropped is less than 5, leave the preceding digit unchanged (i.e., 3.123 → 3.12).
2. If the first digit dropped is greater than 5, increase the preceding digit by 1 (i.e., 3.127 → 3.13).
3. If the first digit dropped is 5, round off to make the preceding digit an even number* (i.e., 4.125 → 4.12; 4.135 → 4.14). The effect of this rule is that, on the average, the retained digit is increased half the time and left unchanged half the time.

Example 6.5 Round off each of the following to 3 significant figures:

a. 6.167 b. 2.132 c. 0.002245 d. 3135

Solution

a. 6.17
b. 2.13
c. Since the digit to be retained, 4, is even, we leave it unchanged, giving 0.00224.
d. Rounding off to an even number gives 3140.

Some calculators automatically round off numbers, or can be made to do so by pressing selected keys. In general, the rounding off is as we have described, except that when the first digit dropped is 5 the number is always rounded up. That is, 4.125 is rounded to 4.13, 4.135 to 4.14, and so on.

6.5 LOGARITHMS, ANTILOGARITHMS

Clearly, the precision of a number will govern the precision of its logarithm. This principle leads to the following general rule:

1. ***IN TAKING THE LOGARITHM OF A NUMBER, RETAIN*** **AFTER THE DECIMAL POINT** ***IN THE LOG AS MANY DIGITS AS THERE ARE SIGNIFICANT FIGURES IN THE NUMBER.***

$\log 3.000 = 0.4771$ $\log 3.000 \times 10^5 = 5.4771$

$\log 3.00 = 0.477$ $\log 3.00 \times 10^3 = 3.477$

$\log 3.0 = 0.48$ $\log 3.0 \times 10^2 = 2.48$

$\log 3 = 0.5$ $\log 3 \times 10^{-4} = 0.5 - 4 = -3.5$

---

*Left-handed people often prefer to round off to odd numbers. It really doesn't matter as long as you're consistent.

Notice that unless the number that precedes the decimal point in the logarithm is zero, there will be more digits in the log than in the number itself. For example, there are 4 significant figures in 3.000×10^5, but there are 5 digits quoted in its logarithm, 5.4771. This is quite reasonable if you think about it. Digits to the left of the decimal point in a log, such as the "5" in 5.4771, serve only to show the power of 10 involved. They are not related in any way to the precision of the number itself.

To find the number corresponding to a given logarithm, we follow an analogous rule:

2. ***IN TAKING AN ANTILOGARITHM, RETAIN AS MANY SIGNIFICANT FIGURES AS THERE ARE DIGITS*** **AFTER THE DECIMAL POINT** ***IN THE LOGARITHM.***

antilog 0.3010 = 2.000	antilog 5.3010 = 2.000×10^5
antilog 0.301 = 2.00	antilog 1.301 = 2.00×10^1
antilog 0.30 = 2.0	antilog 2.30 = 2.0×10^2
antilog 0.3 = 2	antilog (0.3 − 4) = 2×10^{-4}

Example 6.6 Express, to the correct number of significant figures,

a. log 6.19 b. log 6.19×10^{-4}
c. antilog of −0.5231 d. antilog of −2.5231

Solution

a. Using your calculator, you may read: log 6.19 = 0.79169065 . . Following the first rule cited above, you should retain 3 digits to the right of the decimal point. (There are 3 significant figures in 6.19.) Hence,

$$\log 6.19 = 0.792$$

b. $\log 6.19 \times 10^{-4} = -4 + 0.792 = -3.208$
c. Here, the display on your calculator should read 0.2998472 . . Following the second rule cited above, you should retain 4 significant figures. (There are 4 digits to the right of the decimal point in −0.5231). Hence,

$$\text{antilog of } -0.5231 = 0.2998 = 2.998 \times 10^{-1}$$

d. antilog of $-2.5231 = 0.002998 = 2.998 \times 10^{-3}$

PROBLEMS

Observe the rules of significant figures in expressing your answers.

6.1 Give the number of significant figures in
a. 12.82 cm^3
b. 3.19×10^{15} atoms
c. 4.300×10^{-6} cm
d. 0.00641 g
e. 8.2354×10^{-9} m

6.11 Several students determine the concentration of acetic acid in vinegar. The values they report are listed below. How many significant figures are there in each case?
a. 0.1032 mol/ℓ *b.* 0.1030 mol/ℓ
c. 0.0993 mol/ℓ *d.* 1.06×10^{-1} mol/ℓ
e. 1.042×10^{-4} mol/cm^3
f. 0.00010 mol/cm^3
g. 9.2×10^{-5} mol/cm^3

6.2 Give the number of significant figures in

a. 0.0559 g
b. 2.92×10^2 g
c. 4.1 cm^3
d. 0.0002 cm
e. 450 g

6.3 Assuming that all numbers are inexact, carry out the indicated operations:

a. $(2.49 \times 10^{-3})(3.81)$
b. 6.4023×19
c. 7.17/6.2
d. 8.73/5.198

6.4 Assuming that all numbers are inexact, carry out the indicated operations:

a. 0.00481×212
b. $(3.18 \times 10^{-3})(4.2 \times 10^6)$
c. $(6.48 \times 1.92)/5.2$
d. $\dfrac{(8.10 \times 10^7)(4.43 \times 10^{-4})}{6.191 \times 10^2}$

6.5 Given that

$$x = yz/uv$$

where y, z, u, and v are inexact quantities, calculate x if $y = 1.610$, $z = 0.0021$, $u = 1.62 \times 10^{-2}$, and $v = 9.142 \times 10^1$.

6.6 Given that

$$x = yz^2$$

where y and z are inexact quantities, calculate x if

a. $y = 1.082$, $z = 3.1 \times 10^{-3}$
b. $y = 1.69$, $z = 3.24 \times 10^{-2}$

6.12 Consider the element lithium. Among its properties are the following:

atomic mass: 6.941

atomic radius: 0.152 nm

boiling point: 1326°C

charge of ion: +1

ionic radius: 0.060 nm

How many significant figures are there in each case?

6.13 To calibrate a pipet, a student weighs the water which drains from it after it has been filled to the mark. He obtains a mass of 9.9654 g. The density of water is 0.9970 g/cm^3. What is the volume of the pipet (volume = mass/density)?

6.14 When carbon burns in air, 3.664 g of CO_2 is formed for every gram of carbon. What mass of CO_2 is formed from samples of carbon weighing

a. 2.000 g?
b. 6.2 g?

6.15 A sample of gas weighing 1.602 g occupies 224 cm^3 at a pressure of 749 mm Hg and a temperature of 100.0°C. Calculate the molar mass (GMM) of the gas, using the equation

$$\text{GMM} = gRT/PV$$

where g = mass in grams
T = °C + 273.15
P = (pressure in mm Hg)/760
V = volume in cubic centimeters
R = 82.06

6.16 The unit cell of the metal gold is a cube 4.09×10^{-8} cm on an edge.

a. What is the volume, in cubic centimeters, of the unit cell?
b. What is the volume of 6.022×10^{23} unit cells?

6.7 Carry out the indicated operations:

a. 9.10 g + 6.231 g

b. 8.162 cm^3 − 2.39 cm^3

c. 4.30 cm + 29.1 cm + 0.345 cm

d. 6.23 m + 915 cm − 12.7 cm
(note that 1 m = 100 cm)

6.8 Carry out the indicated operations:

a. $(2.02 \times 10^2 \text{ g}) - (9.6 \times 10^1 \text{ g})$

b. $(3.18 \times 10^{-1} \text{ cm}) + (1.6 \times 10^{-2} \text{ cm})$

c. $(6.40 \text{ cm} \times 12.1 \text{ cm}) - 2.19 \text{ cm}^2$

d. $\dfrac{3.18 \text{ g}}{(1.92 \text{ cm})(2.4 \text{ cm})} - 0.17 \text{ g/cm}^2$

6.9 Obtain, to the correct number of significant figures,

a. log 1.602 *b*. log 5.2

c. log 2.18×10^3 *d*. antilog 2.0

e. antilog 0.185 *f*. antilog 3.20

6.10 Given that

$$y = 2 \log x$$

calculate *y* to the correct number of significant figures if

a. $x = 1.614$ *b*. $x = 1.6 \times 10^{-4}$

6.17 What is the total mass of a solution made by adding 12.0 g of sodium chloride, 4.28 g of potassium nitrate, and 1.03 g of potassium chromate to 6.00×10^2 g of water?

6.18 A sample of a certain compound weighing 2.040 g is found by analysis to contain 0.721 g of carbon and 0.050 g of hydrogen. The only other element present is iodine.

a. What is the mass of iodine in the sample?

b. What are the mass percents of the three elements in the compound?

6.19 A student measures the pH of a solution, using instruments of successively greater precision. The values she obtains are

a. 4 *b*. 4.1 *c*. 4.12 *d*. 4.118

Using the definition pH = −log(conc. H^+), calculate the concentration of H^+ corresponding to each measurement.

6.20 The equation relating vapor pressure, *P*, to temperature, *T*, can be written in the form

$$\log P = \frac{-\Delta H}{2.303\ RT} + B$$

where $R = 8.314$ J/K, $T = °\text{C} + 273.15$, ΔH = heat of vaporization in joules, and $B = 5.169$ when *P* is in millimeters of mercury. If the vapor pressure is 12.8 mm Hg at 25.00°C, calculate ΔH.

CHAPTER 7

Algebra: First Degree Equations

Certain practical problems are most readily solved by the use of algebraic equations. To illustrate the approach involved, let us consider a simple case. Suppose you are working in an automobile assembly plant, installing tires as cars come down the line. You have 700 tires on hand; how many cars can be fitted with this number of tires? You might "solve" this problem by setting up the equation

$$4x = 700 \tag{7.1}$$

where x represents the (unknown) number of cars. The "4" takes into account the fact that each automobile requires four tires. This equation is readily solved by dividing both sides by 4. This way, you find that $x = 175$. In other words, 700 tires will outfit 175 cars.

Equation 7.1 is a particularly simple type of algebraic equation. It contains a single unknown, represented by x. Here, the unknown appears in only one term, on one side of the equation. This makes the equation rather easy to solve for the numerical value of x. Sometimes, the unknown may appear on both sides of the equation:

$$5x - 8 = 3x + 12 \tag{7.2}$$

There may be a fraction in the equation:

$$\frac{7x}{3} = 2x - 4 \tag{7.3}$$

The unknown, x, may appear in the denominator of a fraction rather than the numerator:

$$3.0 = \frac{0.63}{x} \tag{7.4}$$

Equations 7.2–7.4 are somewhat more complex than 7.1. However, all four equations have one property in common. They contain a *single unknown, x,* raised to the *first power*. Such equations are often referred to as **first degree equations** in one unknown; the degree of an equation describes the highest power to which an unknown is raised. We will consider a general approach to the solution of such equations in Section 7.1. First degree equations with two unknowns (x, y) are discussed in Section 7.2.

Many problems in general chemistry require solving first degree algebraic equations. Often, we need only substitute numbers into such an equation and solve for an unknown. Occasionally, we may need to combine two equations to obtain the relation we want. Both of these procedures will be covered in Section 7.3.

Chemical equations such as

$$2\ H_2\ (g) + O_2\ (g) \rightarrow 2\ H_2O\ (l) \qquad (7.5)$$

behave, in some ways, like algebraic equations. In particular, the rules used to solve first degree algebraic equations can be applied to chemical equations as well. In Section 7.4, we will look at some of the areas where we can take advantage of this property of chemical equations.

7.1 FIRST DEGREE EQUATIONS IN ONE UNKNOWN

Equations of this type are readily solved by applying a basic principle of algebra. This states that **an equation remains valid if the same operation is performed on both sides.** To be specific, we can:

1. *Add* the same quantity to both sides.
2. *Subtract* the same quantity from both sides.
3. *Multiply* both sides by the same quantity.
4. *Divide* both sides by the same quantity.

Example 7.1 Solve the following equations for x:

a. $2x = 6$ b. $5x - 8 = 3x + 12$ c. $\frac{7x}{3} = 2x - 4$

Solution

a. Divide both sides of the equation by 2:

$$\frac{2x}{2} = \frac{6}{2};\ x = 3$$

b. Several steps are involved:

Add 8 to both sides $5x - 8 + 8 = 3x + 12 + 8;$ $5x = 3x + 20$

Subtract $3x$ from both sides $5x - 3x = 3x - 3x + 20;$ $2x = 20$

Divide both sides by 2 $\frac{2x}{2} = \frac{20}{2};\ x = 10$

c. Multiply both sides by 3 $7x = 6x - 12$

Subtract $6x$ from both sides $x = -12$

Notice the general approach illustrated by Example 7.1. We collect all the terms involving x on one side of the equation and all the numbers on the other side. This approach, combined with the rules cited at the beginning of this section, can be used to solve any first degree equation. This includes equations in which x appears in the denominator of a fraction. Here, it is usually simplest to start by clearing the fractions (Example 7.2).

Example 7.2 Solve the following equations for x:

a. $3.0 = \frac{0.063}{x}$ b. $4 + \frac{1}{x} = \frac{3}{2x}$ c. $\frac{5}{2x - 1} = \frac{2}{x + 4}$

Solution

a. Multiply both sides of the equation by x $3.0x = 0.063$

Divide both sides of the equation by 3.0 $x = 0.021$

b. Here, the simplest approach is to clear the fractions. To do this, we multiply both sides of the equation by $2x$:

$$4(2x) + \frac{2x}{x} = \frac{3(2x)}{2x}$$

Simplifying $8x + 2 = 3$
Now we subtract 2 from both sides and divide by 8:

$$8x = 1; \qquad x = \frac{1}{8}$$

c. Again, we start by clearing fractions. To do this, we multiply both sides by the product of the denominators $(2x - 1)(x + 4)$:

$$\frac{(2x - 1)(x + 4)5}{2x - 1} = \frac{(2x - 1)(x + 4)2}{x + 4}$$

Simplifying $5(x + 4) = 2(2x - 1)$

$5x + 20 = 4x - 2$

Solving $x = -22$

7.2 SIMULTANEOUS FIRST DEGREE EQUATIONS IN TWO UNKNOWNS

Sometimes we have occasion to work with two unknowns rather than one. Suppose, for example, we are dealing with two people, John and Mary, of "unknown" ages x and y. A single equation relating these unknowns cannot be solved for either x or y. If John is three years older than Mary, we can say that

$$x - y = 3 \qquad (7.6)$$

where x is John's age and y is Mary's. Clearly, we cannot solve this equation for the individual ages. Mary could be 20 and John 23, or Mary could be 25 and John 28, or . . . .

To find a single, unique set of values for x and y, we must be given two independent* equations relating these unknowns. Returning to our discussion of John's and Mary's ages, suppose we are told that their total age is 43: then

$$x + y = 43 \qquad (7.7)$$

As we will soon see, the two equations, 7.6 and 7.7 can be used together to obtain individual values of x and y. In other words, by combining these two *simultaneous* relations, we can find the ages of both John and Mary.

A general approach to solving two simultaneous equations is to combine them in such a way as to obtain one equation in a single unknown, x or y. This is then solved in the usual way (Section 7.1). Once x or y is known, the other unknown is readily found by substitution in the original equation. To illustrate this approach, let us apply it to Equations 7.6 and 7.7:

$$x - y = 3 \qquad (7.6)$$

$$x + y = 43 \qquad (7.7)$$

To solve these equations, we could

1. ADD THE EQUATIONS, ELIMINATING y:

$$x - y + (x + y) = 3 + 43$$

$$2x = 46; \qquad x = 23$$

To solve for y, substitute $x = 23$ in Equation 7.6:

$$23 - y = 3; \qquad y = 23 - 3 = 20$$

(The same result would be obtained if you substituted $x = 23$ in Equation 7.7; try it!)

2. SUBTRACT EQUATION 7.6 FROM 7.7, ELIMINATING x:

$$x + y - (x - y) = 43 - 3$$

$$2y = 40; \qquad y = 20$$

To solve for x, substitute $y = 20$ in Equation 7.6:

$$x - 20 = 3; \qquad x = 23$$

Either way, by adding or subtracting equations, we conclude that John is 23 years old and Mary is 20.

Usually, a pair of simultaneous equations is a bit more difficult to solve than the two just considered. In particular, the unknowns usually have different coefficients in

---

*The equations $x - y = 3$ and $2x - 2y = 6$ are *not* independent. The second equation can be obtained from the first by multiplying by 2. These two equations really represent only one relation between x and y. Hence, they can not be solved for unique values of x and y. You can check this by showing that both pairs of numbers listed above ($x = 23$, $y = 20$; $x = 28$, $y = 25$) satisfy *both* equations.

the two equations. Consider, for example,

$$x + 5y = 20 \qquad (7.8)$$

$$3x - 4y = 3 \qquad (7.9)$$

Here, the coefficient of x is 1 in Equation 7.8, as opposed to 3 in Equation 7.9. Similarly, the coefficients of y differ: 5 in Equation 7.8, 4 in Equation 7.9. Hence, we cannot eliminate an unknown simply by adding the two equations or subtracting one from the other. First, we must operate on the equations to make the coefficients of either x or y the same. This can be done by multiplying by appropriate factors. The technique used is illustrated in Example 7.3.

Example 7.3 Solve the simultaneous equations

$$x + 5y = 20 \qquad (7.8)$$

$$3x - 4y = 3 \qquad (7.9)$$

by eliminating

a. x b. y

Solution

a. To eliminate x, we must make the coefficients of this unknown the same in the two equations. This can be done quite simply by multiplying both sides of Equation 7.8 by 3:

$$3x + 15y = 60$$

$$3x - 4y = 3$$

Now we *subtract*, obtaining

$$3x + 15y - (3x - 4y) = 60 - 3$$

$$19y = 57; \qquad y = 3$$

To find x, we substitute $y = 3$ in Equation 7.8:

$$x + 15 = 20; \qquad x = 5$$

b. To eliminate y, we must make its coefficient the same in the two equations. Let us multiply Equation 7.8 by 4 and 7.9 by 5:

$$4x + 20y = 80$$

$$15x - 20y = 15$$

Now we *add*, obtaining

$$4x + 20y + (15x - 20y) = 80 + 15$$

$$19x = 95; \qquad x = 5$$

To find y, we substitute $x = 5$ in Equation 7.8:

$$5 + 5y = 20$$

$$5y = 15; \qquad y = 3$$

The answers are the same as those obtained in part a. This is as it should be, since there is only one set of values of x and y for a given pair of simultaneous equations.

In summary, a pair of simultaneous equations in x and y can be solved in the following way:

1. **MAKE THE COEFFICIENTS OF ONE UNKNOWN,** let us say x, **THE SAME IN THE TWO EQUATIONS.** To do this, multiply one or both equations by appropriate factors.
2. **ADD OR SUBTRACT THE EQUATIONS SO AS TO ELIMINATE THE UNKNOWN (x) REFERRED TO IN (1).** In this way you obtain a single equation in the other unknown (y).
3. **SOLVE THE EQUATION OBTAINED IN (2) FOR THE UNKNOWN (y).**
4. **TO EVALUATE THE SECOND UNKNOWN (x), SUBSTITUTE THE VALUE FOR y OBTAINED IN (3) IN ONE OF THE ORIGINAL EQUATIONS AND SOLVE.**

The approach just described is a general one which can be used for any pair of first degree equations. Indeed, it can be extended to systems of equations containing more than two unknowns. Here, we should point out that **to solve for n unknowns, we must have an equal number, n, of independent equations.** Thus,

1 equation is required to solve for one unknown, x.
2 equations are required to solve for two unknowns, x and y.
3 equations are required to solve for three unknowns, x, y, and z.

7.3 FIRST DEGREE EQUATIONS IN CHEMISTRY

Many problems in general chemistry involve equations containing two or more quantities (e.g., °F and °C; P, V, and T; . . .). Typically, you may be given numerical values for all but one of these quantities. The quantity which remains then becomes the unknown in a first degree algebraic equation. The equation is solved by carrying out one or more of the simple operations described in Section 7.1. Examples 7.4 and 7.5 are typical of this type of problem.

Example 7.4 The relation between temperature expressed in degrees Fahrenheit (°F) and degrees Celsius (°C) is

$$°F = 1.8(°C) + 32°$$

a. Convert 25°C to °F.
b. Convert 98.6°F (normal body temperature) to °C.

Solution

a. Here we can solve for °F by direct substitution:

$$°F = 1.8(25°) + 32° = 77°$$

b. In this case, somewhat more algebra is involved. We first solve the basic equation for the quantity we want, °C:

$$°F - 32° = 1.8(°C) \quad \text{(subtract 32° from both sides)}$$

$$°C = \frac{°F - 32°}{1.8} \quad \text{(divide each side by 1.8)}$$

Then we substitute numbers:

$$°C = \frac{98.6° - 32°}{1.8} = \frac{66.6°}{1.8} = 37.0°$$

Example 7.5 The ideal gas law relates the pressure (P), volume (V), number of moles (n), and Kelvin temperature (T) of a gas:

$$PV = nRT$$

The constant R in this equation has the value 0.0821 $\ell \cdot$ atm/(mol $\cdot$ K). What volume is occupied by 1.60 mol of an ideal gas at 1.20 atm and 300 K?

Solution

We first solve the equation for the quantity we want, V. To do this, we divide both sides by P:

$$V = \frac{nRT}{P}$$

All the quantities on the right side of this equation are given in the statement of the problem. We can obtain V by direct substitution.

$$V = \frac{(1.60 \text{ mol})(0.0821 \frac{\ell \cdot \text{atm}}{\text{mol} \cdot \text{K}})(300 \text{ K})}{1.20 \text{ atm}} = 32.8\ell$$

Sometimes, in general chemistry you will work with first degree equations in which no numbers are involved. An equation may consist entirely of letters standing for different variables. Such an equation is readily "solved" for one variable, following the rules cited in Section 7.1. Example 7.6 illustrates the approach used.

Example 7.6 For a sample of gas in two different states, the following equation applies:

$$\frac{P_1V_1}{T_1} = \frac{P_2V_2}{T_2}$$

where P represents pressure, V is volume, and T is temperature. The subscript 1 refers to the initial state, 2 to the final state. Solve this equation for T_2.

Solution

The approach here is similar to that followed in Example 7.2. Let us start by clearing fractions. To do this, we multiply both sides of the equation by the product, $T_2 \times T_1$:

$$T_2T_1\left(\frac{P_1V_1}{T_1}\right) = T_2T_1\left(\frac{P_2V_2}{T_2}\right)$$

$$T_2P_1V_1 = T_1P_2V_2$$

To solve for T_2, all we need do now is divide both sides by P_1V_1:

$$\frac{T_2P_1V_1}{P_1V_1} = \frac{T_1P_2V_2}{P_1V_1}$$

$$T_2 = \frac{T_1P_2V_2}{P_1V_1}$$

Occasionally, to solve a problem in chemistry it is necessary to modify a basic equation such as the ideal gas law. Often this can be done by making a simple substitution and then applying the basic principles of algebra (Example 7.7).

Example 7.7 Obtain an expression for the molar mass GMM of a gas in terms of its pressure P, volume V, temperature T, and mass in grams g. Start with the ideal gas law referred to in Example 7.5:

$$PV = nRT$$

Solution

Before you can work this problem, you have to stop and think for a moment about the meaning of the terms in the ideal gas law. In particular, you must consider carefully what is meant by the number of moles, n. *The number of moles, n, can be found by dividing the number of grams, g, by the molar mass,* GMM. That is,

$$n = \frac{g}{\text{GMM}}$$

Once you recognize this relation, the algebra is straightforward. Substitute g/GMM for its equivalent, n, in the ideal gas law:

$$PV = \frac{gRT}{\text{GMM}}$$

Now all that remains is to solve this equation for GMM:

$$PV(\text{GMM}) = gRT \qquad \text{(multiply both sides by GMM)}$$

$$\text{GMM} = \frac{gRT}{PV} \qquad \text{(divide both sides by } PV\text{)}$$

7.4 CHEMICAL EQUATIONS AS ALGEBRAIC EQUATIONS

In many respects, balanced chemical equations can be treated as if they were algebraic equations. Some of the rules discussed in Section 7.1 can be applied usefully to chemical equations. In particular:

1. **A CHEMICAL EQUATION REMAINS VALID IF BOTH SIDES ARE MULTIPLIED BY THE SAME NUMBER.** Thus, the equations

$$H_2\text{ (g)} + \tfrac{1}{2}\, O_2\text{ (g)} \rightarrow H_2O\text{ (l)}$$

$$2\, H_2\text{ (g)} + O_2\text{ (g)} \rightarrow 2\, H_2O\text{ (l)}$$

are both valid ways of describing the reaction between hydrogen and oxygen. The coefficients (1, $\frac{1}{2}$, 1 in the first equation; 2, 1, 2 in the second) tell us the relative numbers of moles of reactants and products.

2. **TWO CHEMICAL EQUATIONS CAN BE ADDED ALGEBRAICALLY TO GIVE A THIRD, VALID EQUATION.** Consider, for example, the equations

$$C\text{ (s)} + \tfrac{1}{2}\, O_2\text{ (g)} \rightarrow CO\text{ (g)}$$

$$CO\text{ (g)} + \tfrac{1}{2}\, O_2\text{ (g)} \rightarrow CO_2\text{ (g)}$$

Suppose we treat these equations as algebraic equations and add them. Then we get

$$C\ (s) + \tfrac{1}{2}\ O_2\ (g) + CO\ (g) + \tfrac{1}{2}\ O_2\ (g) \rightarrow CO\ (g) + CO_2\ (g)$$

To simplify, we subtract a CO term from both sides and combine the two terms involving O_2. When we do this, the result is

$$C\ (s) + O_2\ (g) \rightarrow CO_2\ (g)$$

This is a valid, balanced equation for the reaction of carbon with oxygen to give carbon dioxide, CO_2.

These properties of chemical equations are useful in several areas. In the remainder of this section we will consider two such areas: balancing redox equations and thermochemistry (Hess' Law).

Balancing Redox Equations

A common method of balancing oxidation-reduction equations uses what are known as "half-equations." In one type of half-equation, referred to as oxidation, electrons (e^-) appear on the right. In the other, called reduction, electrons are written on the left. Typical half-equations are

oxidation $\quad 2\ Cl^-\ (aq) \rightarrow Cl_2\ (g) + 2\ e^- \quad$ (7.10)

reduction $\quad MnO_4^-\ (aq) + 8\ H^+\ (aq) + 5\ e^- \rightarrow Mn^{2+}\ (aq) + 4\ H_2O \quad$ (7.11)

To obtain a balanced equation for an overall reaction of oxidation and reduction, we combine an oxidation half-equation with a reduction half-equation. The chemical principle involved is a simple one. *There must be no electrons in the overall equation.* Electrons are "conserved" in a chemical reaction. That is, they are neither created nor destroyed.

The approach used to combine half-equations resembles that described in Example 7.3. We want to eliminate electrons (e^-) between the two equations. To do this, we must make the coefficients of e^- the same in the two half-equations. This is accomplished by multiplying one or both equations by appropriate numbers. Then, the equations are added to "cancel" electrons, giving a balanced overall equation.

Let us apply this approach to combine half-equations 7.10 and 7.11 above. We multiply Equation 7.10 by 5 and Equation 7.11 by 2:

$$10\ Cl^-\ (aq) \rightarrow 5\ Cl_2\ (g) + 10\ e^-$$

$$2\ MnO_4^-\ (aq) + 16\ H^+\ (aq) + 10\ e^- \rightarrow 2\ Mn^{2+}\ (aq) + 8\ H_2O$$

Now we add equations:

$$10\ Cl^-\ (aq) + 2\ MnO_4^-\ (aq) + 16\ H^+\ (aq) + 10\ e^- \rightarrow$$

$$5\ Cl_2\ (g) + 10\ e^- + 2\ Mn^{2+}\ (aq) + 8\ H_2O$$

Simplifying by subtracting 10 e^- from both sides of the equation,

$$10\ Cl^-\ (aq) + 2\ MnO_4^-\ (aq) + 16\ H^+\ (aq) \rightarrow 5\ Cl_2\ (g) + 2\ Mn^{2+}\ (aq) + 8\ H_2O \qquad (7.12)$$

This is the balanced equation for the oxidation-reduction reaction between Cl^- ions and MnO_4^- ions.

Example 7.8 Combine the two half-equations

oxidation $\quad NO\ (g) + 2\ H_2O \rightarrow NO_3^-\ (aq) + 4\ H^+\ (aq) + 3\ e^-$

reduction $\quad O_2\ (g) + 4\ H^+\ (aq) + 4\ e^- \rightarrow 2\ H_2O$

to give a balanced redox equation for the reaction of NO with O_2 in water solution.

Solution

To eliminate electrons, we start by multiplying the first equation by 4 and the second equation by 3:

$$4\ NO\ (g) + 8\ H_2O \rightarrow 4\ NO_3^-\ (aq) + 16\ H^+\ (aq) + 12\ e^-$$

$$3\ O_2\ (g) + 12\ H^+\ (aq) + 12\ e^- \rightarrow 6\ H_2O$$

Simple addition gives

$$4\ NO\ (g) + 8\ H_2O + 3\ O_2\ (g) + 12\ H^+\ (aq) + 12\ e^- \rightarrow$$
$$4\ NO_3^-\ (aq) + 16\ H^+\ (aq) + 12\ e^- + 6\ H_2O$$

We can simplify by subtracting 12 e^-, 6 H_2O, and 12 H^+ from each side of this equation:

$$4\ NO\ (g) + 2\ H_2O + 3\ O_2\ (g) \rightarrow 4\ NO_3^-\ (aq) + 4\ H^+\ (aq)$$

Thermochemistry (Hess' Law)

A thermochemical equation is one in which the enthalpy change, ΔH, is specified. Ordinarily this is done by quoting, at the right of the equation, the value of ΔH in kilojoules.

$$C\ (s) + \tfrac{1}{2}\ O_2\ (g) \rightarrow CO\ (g) \qquad \Delta H = -110.5\ kJ \qquad (7.13)$$

This equation tells us that, when one mole of CO is formed from the elements, ΔH is -110.5 kJ. This means that 110.5 kJ of heat is evolved in the reaction. A negative value of ΔH corresponds to a reaction in which heat is evolved. If heat is absorbed in a reaction, the sign of ΔH is positive.

There are three important rules that apply to thermochemical equations:

1. **IF THE COEFFICIENTS OF THE SPECIES IN THE EQUATION ARE MULTIPLIED BY A NUMBER, *n*, THE VALUE OF ΔH IS MULTIPLIED BY THAT SAME NUMBER, *n*.**

Thus, if we multiply Equation 7.13 by 2,

$$2\ C\ (s) + O_2\ (g) \rightarrow 2\ CO\ (g) \qquad \Delta H = -221.0\ kJ$$

2. IF THE DIRECTION OF AN EQUATION IS REVERSED, THE SIGN OF ΔH CHANGES. Again, referring to Equation 7.13,

$$CO\ (g) \rightarrow C\ (s) + \tfrac{1}{2}\ O_2\ (g) \qquad \Delta H = +110.5\ kJ$$

3. IF AN EQUATION (3) IS THE SUM OF TWO OTHER EQUATIONS (1 AND 2), THEN $\Delta H_3 = \Delta H_1 + \Delta H_2$. To illustrate the use of this rule, known as **Hess' Law**, consider the two equations:

$$(1)\quad C\ (s) + \tfrac{1}{2}\ O_2\ (g) \rightarrow CO\ (g) \qquad \Delta H_1 = -110.5\ kJ$$

$$(2)\quad CO\ (g) + \tfrac{1}{2}\ O_2\ (g) \rightarrow CO_2\ (g) \qquad \Delta H_2 = -283.0\ kJ$$

Adding the two chemical equations algebraically, we obtain

$$(3)\quad C\ (s) + O_2\ (g) \rightarrow CO_2\ (g)$$

Applying Hess' Law,

$$\Delta H_3 = \Delta H_1 + \Delta H_2 = -110.5\ kJ - 283.0\ kJ = -393.5\ kJ$$

In other words, the thermochemical equation for the formation of CO_2 is:

$$C\ (s) + O_2\ (g) \rightarrow CO_2\ (g) \qquad \Delta H = -393.5\ kJ$$

Hess' Law is very useful in thermochemistry, but for it to be effective, you must use the algebraic properties of chemical equations. Typically, you are given ΔH for two or more equations and asked to determine ΔH for a different, but related, equation. To do this, often you need to apply rules (1) and (2), listed above, as well as Hess' Law.

To illustrate a typical Hess' Law calculation, suppose you are given the thermochemical equations

$$MnO\ (s) + \tfrac{1}{2}\ O_2\ (g) \rightarrow MnO_2\ (s) \qquad \Delta H = -134.8\ kJ \qquad (7.14)$$

$$MnO_2\ (s) + Mn\ (s) \rightarrow 2\ MnO\ (s) \qquad \Delta H = -250.1\ kJ \qquad (7.15)$$

You are asked to calculate ΔH for

$$Mn\ (s) + O_2\ (g) \rightarrow MnO_2\ (s) \qquad \Delta H = ? \qquad (7.16)$$

Clearly, you cannot apply Hess' Law directly; Equations 7.14 and 7.15 do *not* add to 7.16. Instead, you must modify one or both of these equations before adding. To decide what to do, it is helpful to note that MnO, which appears in both 7.14 and

7.15, is *not* present in the final equation. This suggests that you should work to obtain two equations which, when added, will eliminate MnO. To do this you need only multiply Equation 7.14 by 2 and add to 7.15:

$$2\ MnO\ (s) + O_2\ (g) \rightarrow 2\ MnO_2\ (s) \qquad \Delta H = -269.6\ kJ$$

$$MnO_2\ (s) + Mn\ (s) \rightarrow 2\ MnO\ (s) \qquad \Delta H = -250.1\ kJ$$

(Note that when Equation 7.14 is multiplied by 2, ΔH is also multiplied by 2.) Adding,

$$2\ MnO\ (s) + O_2\ (g) + MnO_2\ (s) + Mn\ (s) \rightarrow 2\ MnO_2\ (s) + 2\ MnO\ (s)$$

$$\Delta H = -269.6\ kJ - 250.1\ kJ = -519.7\ kJ$$

Simplifying by subtracting like terms ($2\ MnO + MnO_2$) from both sides,

$$Mn\ (s) + O_2\ (g) \rightarrow MnO_2\ (s) \qquad \Delta H = -519.7\ kJ$$

We conclude that ΔH for Equation 7.16 is -519.7 kJ.

The example we have just gone through suggests a general approach for Hess' Law calculations. Your goal should be to *eliminate species which do not appear in the final equation*. To do this, work with the original equations to get an equal number of moles of that species on the left of one equation and the right of the other. Then, apply Hess' Law. If all has gone well, you should obtain the desired equation and its corresponding ΔH.

Example 7.9 Given the thermochemical equations

$$2\ CuO\ (s) \rightarrow Cu_2O\ (s) + \tfrac{1}{2}\ O_2\ (g) \qquad \Delta H = +143.7\ kJ$$

$$CuO\ (s) + Cu\ (s) \rightarrow Cu_2O\ (s) \qquad \Delta H = -11.5\ kJ$$

obtain ΔH for

$$Cu\ (s) + \tfrac{1}{2}\ O_2\ (g) \rightarrow CuO\ (s) \qquad \Delta H = ?$$

Solution

Clearly, the species we want to eliminate is Cu_2O. We can accomplish this by reversing the first equation (remembering to change the sign of ΔH) and then adding

$$Cu_2O\ (s) + \tfrac{1}{2}\ O_2\ (g) \rightarrow 2\ CuO\ (s) \qquad \Delta H = -143.7\ kJ$$

$$CuO\ (s) + Cu\ (s) \rightarrow Cu_2O\ (s) \qquad \Delta H = -11.5\ kJ$$

Adding $\quad Cu_2O\ (s) + \tfrac{1}{2}\ O_2\ (g) + CuO\ (s) + Cu\ (s) \rightarrow 2\ CuO\ (s) + Cu_2O\ (s)$

$$\Delta H = -143.7\ kJ - 11.5\ kJ = -155.2\ kJ$$

Simplifying by subtracting like terms (Cu_2O, CuO) from both sides,

$$Cu\ (s) + \tfrac{1}{2}\ O_2\ (g) \rightarrow CuO(s) \qquad \Delta H = -155.2\ kJ$$

Sometimes the calculations involved in a Hess' Law problem are a bit more complex than those we have shown. In particular, more than one species may have to be eliminated (Example 7.10).

Example 7.10 Given the thermochemical equations

$$\begin{array}{llr} (1)\ C_3H_8\ (g) + 5\ O_2\ (g) \rightarrow 3\ CO_2\ (g) + 4\ H_2O\ (l) & \Delta H = & -2219.9\ kJ \\ (2)\ \qquad C\ (s) + O_2\ (g) \rightarrow CO_2\ (g) & \Delta H = & -\ 393.5\ kJ \\ (3)\ \quad 2\ H_2\ (g) + O_2\ (g) \rightarrow 2\ H_2O\ (l) & \Delta H = & -\ 571.6\ kJ \end{array}$$

calculate ΔH for the reaction

$$3\ C\ (s) + 4\ H_2\ (g) \rightarrow C_3H_8\ (g) \qquad \Delta H = ?$$

Solution

Here, we need to eliminate three species: CO_2, H_2O, and O_2. It is not at all obvious what to do with the O_2, since it appears in all three equations. So, let's concentrate on the CO_2 and H_2O and worry about the O_2 later, if we have to. To eliminate CO_2 and H_2O, we should

— reverse Equation 1 to get 3 CO_2 and 4 H_2O on the left
— multiply Equation 2 by 3 to get 3 CO_2 on the right
— multiply Equation 3 by 2 to get 4 H_2O on the right

Thus, we have

$$\begin{array}{lll} 3\ CO_2\ (g) + 4\ H_2O\ (l) \rightarrow C_3H_8\ (g) + 5\ O_2\ (g) & \Delta H = & +2219.9\ kJ \\ 3\ C\ (s) + 3\ O_2\ (g) \rightarrow 3\ CO_2\ (g) & \Delta H = & -1180.5\ kJ \\ 4\ H_2\ (g) + 2\ O_2\ (g) \rightarrow 4\ H_2O\ (l) & \Delta H = & -1143.2\ kJ \\ \hline 3\ C\ (s) + 4\ H_2\ (g) \rightarrow C_3H_8\ (g) & \Delta H = & -103.8\ kJ \end{array}$$

Notice that, in getting rid of CO_2 and H_2O, we eliminated O_2 as well.

Reviewing Examples 7.8–7.10, you will note certain similarities in the approaches followed. In each case, we sought to eliminate species which did not belong in the final equation. In Example 7.8, the species eliminated was the electron; in 7.9 and 7.10, the species were chemical substances (Cu_2O, CO_2, H_2O, O_2). You may also notice a resemblance between these examples and Example 7.3 in Section 7.2. There, you will recall, we operated on two equations to eliminate a variable. The method we used to solve simultaneous algebraic equations is similar to the approach followed in this section with chemical equations.

PROBLEMS

7.1 Solve each of the following equations for x.

a. $4.0x = 6.0 \times 10^{-4}$

b. $\dfrac{2x - 8}{6} = 3 - 4x$

c. $x = 2y - 8$, when $y = 6$; $y = -4$

7.11 Consider the equation

$$T = t + 273$$

where T is the Kelvin temperature and t the temperature in degrees Celsius. Calculate

a. T when $t = -106$

b. t when $T = 319$

c. the Kelvin Temperature when $t = 0.400\ T$

7.2 Solve each of the following equations for x.

a. $6.0x = (2.0 \times 10^{-2})(1.8 \times 10^{3})$

b. $(12.6)(0.0182) = x(1.4 \times 10^{-8})$

c. $x(1.6 \times 10^{-2}) = \frac{9.8 \times 10^{4}}{1.62}$

7.3 Consider the equation

$$10.0\ y = \frac{6.00\ x}{1 - 18.0\ x}$$

a. Solve this equation for x.

b. Find the value of x when $y = 1.00$; $y = 0.100$.

7.4 Consider the equation

$$y = \frac{1.5/x}{30\ z}$$

a. Solve this equation for x, in terms of y and z.

b. Find x when $y = 1$, $z = 1$; when $y = 2$, $z = 5$.

7.5 Consider the equation

$$xy = \frac{uv}{z}$$

Solve this equation for

a. x *b*. u *c*. z

7.6 Combine the equations

$$x = \frac{y}{z} \quad \text{and} \quad y = \frac{uv}{w}$$

to obtain an equation for z, in terms of x, u, v, and w.

7.12 Consider the ideal gas law

$$PV = nRT$$

where P is the pressure, V is the volume, n is the number of moles, T is the Kelvin temperature, and $R = 0.0821\ell \cdot \text{atm/(mol} \cdot \text{K)}$. Calculate

a. V when $n = 1.00$ mol, $T = 300$ K, $P = 0.952$ atm

b. T when $n = 0.500$ mol, $P = 1.961$ atm, $V = 29.6\ell$

7.13 The molarity, c, of a solution is related to the molality, m, by the equation

$$c = \frac{md}{1 + mM_2/1000}$$

where d is the density (g/cm^3) and M_2 is the molar mass of the solute.

a. Solve this equation for m.

b. Calculate the molality of a 0.1000 molar solution which has a density of 1.004 g/cm^3 and contains a solute of molar mass 184.

7.14 The following relation can be used to determine the molar mass of a solute dissolved in water:

$$-t_f = \frac{1.86 \times g_2/M_2}{g_1/1000}$$

where t_f is the freezing point of the solution, g_2 is the mass in grams and M_2 the molar mass of the solute, and g_1 is the mass in grams of water. A solution of 1.00 g of a certain solute in 20.0 g of water freezes at −0.423°C. Calculate the molar mass of the solute.

7.15 Consider the equation referred to in Example 7.6. Solve this equation for

a. V_2 *b*. V_1 *c*. P_2

d. P_1 *e*. T_1

7.16 Using the equation obtained in Example 7.7

$$\text{GMM} = \frac{gRT}{PV}$$

obtain an expression for the density of an ideal gas (density = mass/volume).

7.7 Solve the following sets of simultaneous equations for x and y:

a. $3x - 4y = 16$

$2x + 5y = 26$

b. $0.0556x + 0.0153y = 0.00709$

$x + y = 0.200$

7.8 Consider the equations

$$x + 2y = 3z$$

$$z + x = 6y$$

Combine these equations to obtain a single equation in which z does not appear.

7.9 Consider the equations

$$2x + y = 6z$$

$$x - 3y = 4z$$

Show that these equations can be combined (after multiplying by the proper factors) to give

$$4x + 2y = 3x - 9y$$

7.10 Consider the equations

$$2x + y = 2z$$

$$u + 2y = 2z$$

Show that these equations can be combined to give

$$u + y = 2x$$

7.17 The atomic mass of chlorine can be calculated from the equation

$$\text{A.M.} = x(36.97) + y(34.97)$$

where x is the fraction of the Cl-37 isotope and y is the fraction of the Cl-35 isotope.

a. The fractions of the two isotopes must add to unity. Knowing that, set up a second equation relating x to y.

b. Solve the two equations (i.e., the one stated and the one obtained in part a) for x and y, taking A.M. Cl = 35.45.

7.18 When zinc reacts with dilute nitric acid, the two half-equations are

oxidation $\quad Zn\ (s) \rightarrow Zn^{2+}\ (aq) + 2\ e^-$

reduction $\quad NO_3^-\ (aq) + 10\ H^+\ (aq) + 8\ e^- \rightarrow NH_4^+\ (aq) + 3\ H_2O$

Using the principle described in Example 7.8, write a balanced, overall equation for the reaction.

7.19 Consider the thermochemical equations

$$2\ H_2O_2\ (l) \rightarrow 2\ H_2O\ (l) + O_2\ (g); \quad \Delta H = +196.4\ \text{kJ}$$

$$H_2O\ (l) \rightarrow H_2\ (g) + \tfrac{1}{2}O_2\ (g); \quad \Delta H = +285.8\ \text{kJ}$$

Using Hess' Law (Section 7.4), calculate ΔH for the reaction

$$2\ H_2O_2\ (l) \rightarrow 2\ H_2\ (g) + 2\ O_2\ (g)$$

7.20 Consider the thermochemical equations

$$2NO\ (g) + O_2\ (g) \rightarrow 2\ NO_2\ (g); \quad \Delta H = -113.0\ \text{kJ}$$

$$N_2\ (g) + 2\ O_2\ (g) \rightarrow 2\ NO_2\ (g); \quad \Delta H = +69.8\ \text{kJ}$$

Use Hess' Law to obtain ΔH for the reaction

$$N_2\ (g) + O_2\ (g) \rightarrow 2\ NO\ (g)$$

CHAPTER

8

Algebra: Second and Higher Degree Equations

In Chapter 7 we considered first degree algebraic equations, where the variables appear only to the first power. In this chapter, we will examine higher degree equations in a single variable, x. For the most part, we will concentrate on second degree equations, commonly called *quadratic* equations, which contain an x^2 term.

Certain quadratic or higher degree equations can be solved quite simply by reduction to a first degree equation (Section 8.1). More generally, quadratic equations involving terms in x^2 and x are solved by using the quadratic formula (Section 8.2). In chemistry, second and higher degree equations arise most often in calculations involving chemical equilibria (Section 8.3). It is important that you be able to set up such equations as well as solve them. Quadratic equations which arise in chemistry can sometimes be solved by reducing them to a first degree equation. More often, you will need to use the quadratic formula or an approximation method that will be discussed in Section 8.4.

8.1 EQUATIONS WHICH CAN BE REDUCED TO FIRST DEGREE

Consider the general equation

$$x^n = a \tag{8.1}$$

where n is an integer greater than 1 (i.e., $n = 2, 3, \ldots$) and a is a number (e.g., 4, $3 \times 10^{-6}, \ldots$). This equation can be reduced to a first degree equation and solved by taking the nth root of both sides:

$$x = (a)^{1/n} \tag{8.2}$$

As an example, consider the equation

$$x^2 = 2.0 \times 10^{-4}$$

To solve this equation, we take the square root of both sides. Using your calculator, you should find that the square root of 2.0×10^{-4} is 1.4×10^{-2}. The answer is

$$x = \pm 1.4 \times 10^{-2}$$

(Note that both 1.4×10^{-2} and -1.4×10^{-2} give 2.0×10^{-4} when squared. Your calculator gives only the positive root.)

The same approach can be used if n is a fraction such as $\frac{1}{2}$ or $\frac{1}{3}$. Consider, for example, the equation

$$x^{1/2} = 3.0 \times 10^{-2}$$

To solve this equation, square both sides:

$$x = (3.0 \times 10^{-2})^2 = 9.0 \times 10^{-4}$$

Example 8.1 Solve the following equations for x:

a. $x^3 = 2.6 \times 10^{-6}$ b. $2x^2 = 3.2 \times 10^3$ c. $x^{1/3} = 1.6$

Solution

a. Extracting the cube root of both sides

$$x = (2.6 \times 10^{-6})^{1/3} = 1.4 \times 10^{-2}$$

(Extracting roots using a calculator is discussed in Chapter 1.)

b. We start by solving for x^2:

$$x^2 = 1.6 \times 10^3$$

Then we extract the square root of both sides:

$$x = \pm 40$$

c. Cubing both sides,

$$x = (1.6)^3 = 4.1$$

The approach shown above can be extended to quadratic equations of a more complex form than Equation 8.1. Consider, for example, the equation

$$(x - 6.0)^2 = 24$$

Here, as before, we can reduce to a first degree equation by taking the square root of both sides:

$$x - 6.0 = (24)^{1/2} = +4.9 \text{ or } -4.9$$

Solving this equation for x,

$$x = 6.0 + 4.9 = 10.9$$

or $$x = 6.0 - 4.9 = 1.1$$

Either of these two values of x satisfies the equation we started with, $(x - 6.0)^2 = 24$.

Example 8.2 Solve the following equation for x:

$$\frac{x^2}{(1.20 - 2x)^2} = 0.010$$

Solution

Note that the left side of this equation is a "perfect square." That is, we can extract its square root exactly. Taking the square root of both sides,

$$\frac{x}{1.20 - 2x} = (0.010)^{1/2} = \pm 0.10$$

This equation can now be solved for x in the usual way. We get two different values for x, depending upon whether we use the positive root, +0.10, or the negative root, −0.10

$+0.10$	-0.10
$x = 0.12 - 0.20x$	$x = -0.12 + 0.20x$
$1.20x = 0.12$	$0.80x = -0.12$
$x = 0.10$	$x = -0.15$

Either of these values of x satisfies the original equation. To show that this is the case, note that

$$\frac{(0.10)^2}{(1.00)^2} = 0.01; \qquad \frac{(-0.15)^2}{(1.50)^2} = 0.010$$

Many of the algebraic equations that arise in a study of chemical equilibria (Section 8.3) are of the type just discussed. They can be solved very simply by extracting the square root of both sides and solving the first degree equation obtained. Students often fail to recognize this and apply more complex methods, wasting a great deal of time in the process. Whenever you are confronted with a quadratic equation, you should first check to see if perchance the terms involving x form a "perfect square." If they do, you can apply the simple procedure illustrated in Example 8.2 to find x.

8.2 SOLVING SECOND DEGREE EQUATIONS BY THE QUADRATIC FORMULA

The method illustrated in Example 8.2 is not applicable to most second degree equations. Consider, for example, the equation

$$\frac{x^2}{1 - x} = 4$$

The left side of this equation is not a "perfect square." Clearly, we cannot solve for x by extracting the square root of both sides of the equation.

Any second degree equation in one unknown can be solved by applying the so-called "quadratic formula." To use this approach, we rewrite the equation, if necessary, to get it in the form

$$ax^2 + bx + c = 0 \tag{8.3}$$

where a, b, and c are numbers (a is the coefficient of the x^2 term; b is the coefficient of the x term). This equation has two roots; that is, there are two values of x that satisfy it. To find these roots, we use a formula derived by a method which we will not consider here. The quadratic formula is

$$x = \frac{-b \pm \sqrt[2]{b^2 - 4ac}}{2a} \tag{8.4}$$

The $\pm$ sign indicates that we are to consider both the positive and negative square roots of $(b^2 - 4ac)$, leading to two different values of x.

To illustrate how the quadratic formula is applied, let us use it to find the two values of x that satisfy the equation

$$\frac{x^2}{1 - x} = 4$$

We first rewrite this equation to get it in the proper form:

$$x^2 = 4 - 4x$$

$$x^2 + 4x - 4 = 0$$

Comparing the equation just written to Equation 8.3, we deduce that

$$a = 1; \qquad b = 4; \qquad c = -4$$

Consequently

$$x = \frac{-4 \pm \sqrt[2]{16 + 16}}{2} = \frac{-4 \pm \sqrt[2]{32}}{2}$$

The square root of 32 is 5.66. So

$$x = \frac{-4 \pm 5.66}{2} = \frac{1.66}{2} \quad \text{or} \quad \frac{-9.66}{2}$$

$$x = 0.83 \text{ or } -4.83$$

Sometimes the arithmetic involved in using the quadratic formula is a bit more complex than that shown above. Example 8.3, which is typical of a type of quadratic equation that arises often in chemistry, illustrates this point.

Example 8.3 Using the quadratic formula, solve the equation

$$\frac{x^2}{0.100 - x} = 1.80 \times 10^{-5}$$

Solution

Rearranging the equation to get it in standard form,

$$x^2 = 1.80 \times 10^{-6} - 1.80 \times 10^{-5}x$$

$$x^2 + 1.80 \times 10^{-5}x - 1.80 \times 10^{-6} = 0$$

$$a = 1; \qquad b = 1.80 \times 10^{-5}; \qquad c = -1.80 \times 10^{-6}$$

Hence

$$x = \frac{-1.80 \times 10^{-5} \pm \sqrt[2]{(3.24 \times 10^{-10}) + 7.20 \times 10^{-6}}}{2}$$

To obtain the square root, we note that $3.24 \times 10^{-10} = 0.000324 \times 10^{-6}$. Hence

$$3.24 \times 10^{-10} + 7.20 \times 10^{-6} = 0.000324 \times 10^{-6} + 7.20 \times 10^{-6}$$

$$\approx 7.20 \times 10^{-6}$$

$$x = \frac{-1.80 \times 10^{-5} \pm \sqrt[2]{7.20 \times 10^{-6}}}{2}$$

$$= \frac{-1.80 \times 10^{-5} \pm 2.68 \times 10^{-3}}{2}$$

$$= \frac{-0.0180 \times 10^{-3} \pm 2.68 \times 10^{-3}}{2}$$

$$= \frac{2.66 \times 10^{-3}}{2} \quad \text{or} \quad \frac{-2.70 \times 10^{-3}}{2}$$

$$x = 1.33 \times 10^{-3} \quad \text{or} \quad -1.35 \times 10^{-3}$$

As we have pointed out, every second degree equation in a single variable has two solutions. That is, there are two different values of x which satisfy the equation. In the "real world" of chemistry, one of these answers can ordinarily be discarded as being unreasonable. Only one answer will make sense in the context of the problem being solved.

8.3 SECOND AND HIGHER DEGREE EQUATIONS IN CHEMICAL EQUILIBRIA

In general chemistry, algebraic equations of the type discussed in this chapter arise most often in problems dealing with chemical equilibria. Before discussing these problems, we will consider briefly the concept of the **equilibrium constant.** A more complete discussion of this quantity can be found in your chemistry text.

For a chemical system at equilibrium, there is an algebraic equation that relates the concentrations of reactants and products. Consider the general system

$$aA + bB \rightleftharpoons cC + dD$$

Here, A, B, C, and D are species in solution, usually a gaseous or aqueous solution. The small letters a, b, c, d represent the coefficients that appear in the balanced chemical equation. For this system, the following algebraic equation applies:

$$\frac{[C]^c \times [D]^d}{[A]^a \times [B]^b} = K_c$$

In this equation, the brackets represent equilibrium concentrations, usually in moles per liter. The symbol K_c stands for the equilibrium constant, a number which is characteristic of a particular reaction at a given temperature.

To illustrate what this statement means, let us consider an equilibrium involving three species, HI, H_2, and I_2. The balanced equation used to describe this system can be written

$$2HI\ (g) \rightleftharpoons H_2\ (g) + I_2\ (g)$$

For this equation, the expression for the equilibrium constant is

$$\frac{[H_2] \times [I_2]}{[HI]^2} = K_c$$

The quantity K_c has a fixed value at a given temperature. It is, for example, 0.016 at 520°C. Regardless of the initial concentrations of HI, H_2, or I_2, reaction will occur until the ratio (conc. H_2)(conc. I_2)/(conc. $HI)^2$ is 0.016 at this temperature. When that ratio is reached, equilibrium is established and no further changes in concentration occur.

Other examples of equilibrium constant expressions are given in Table 8.1. Note that in each case the concentrations of products (right side of equation) appear in the numerator. Those of reactants (left side of equation) appear in the denominator. The power to which each concentration is raised is given by the coefficient of the species in the balanced chemical equation.

Table 8.1 EXAMPLES OF EQUILIBRIUM CONSTANT EXPRESSIONS

	Reaction	Expression
1.	$2HI(g) \rightleftharpoons H_2(g) + I_2(g)$	$K_c = \frac{[H_2] \times [I_2]}{[HI]^2}$
2.	$N_2(g) + O_2(g) \rightleftharpoons 2NO(g)$	$K_c = \frac{[NO]^2}{[N_2] \times [O_2]}$
3.	$N_2(g) + 3H_2(g) \rightleftharpoons 2NH_3(g)$	$K_c = \frac{[NH_3]^2}{[N_2] \times [H_2]^3}$
4.	$HC_2H_3O_2(aq) \rightleftharpoons H^+(aq) + C_2H_3O_2^-(aq)$	$K_a^* = \frac{[H^+] \times [C_2H_3O_2^-]}{[HC_2H_3O_2]}$

*The equilibrium constant for this type of reaction, the dissociation of a weak acid into ions, is referred to as the "dissociation constant" or "ionization constant" of the acid and is given the special symbol K_a.

Students commonly find problems involving equilibrium constants among the most difficult of those in the beginning course. In part, this is because such problems come in a wide variety of different forms. Here we will concentrate upon one aspect that is common to nearly all equilibrium problems. The operation that we will examine can be stated as follows:

Given the numerical value of K_c and the initial concentrations of all species, set up an algebraic equation in one unknown that can be solved to yield the equilibrium concentrations of all species.

In some cases (Example 8.4), the algebraic equation obtained can be solved by reducing it to a first degree equation. More often, this is not possible (Examples 8.5 and 8.6). It is necessary to use either the quadratic formula (Section 8.2) or an approximation method (Section 8.4).

Example 8.4 The equilibrium constant for the system:

$$2HI\ (g) \rightleftharpoons H_2\ (g) + I_2\ (g)$$

is 0.010 at a particular temperature. The initial concentrations are as follows:

$$HI = 1.20\ \text{mol}/\ell, \qquad H_2 = 0, \qquad I_2 = 0$$

Set up an algebraic equation which relates the equilibrium concentrations of all species. Solve it to obtain $[H_2]$, $[I_2]$, and $[HI]$.

Solution

The equilibrium constant expression is

$$K_c = 0.010 = \frac{[H_2] \times [I_2]}{[HI]^2}$$

It seems reasonable to take the unknown, x, to be the equilibrium concentration of H_2. To express $[I_2]$ and $[HI]$ in terms of x, we reason as follows:

1. The chemical equation tells us that one mole of I_2 is formed for every mole of H_2. Since the concentration of H_2 increased by x moles per liter, that of I_2 must have increased by the same amount, x. Noting that the initial concentration of I_2 is zero, it follows that $[I_2] = x$.
2. According to the chemical equation, two moles of HI are consumed for every mole of H_2 formed. Hence, if we form x moles per liter of H_2, we must use up $2x$ moles per liter of HI. Remember that we started with 1.20 moles per liter of HI. Hence

$$[HI] = (1.20 - 2x)\ \text{mol}/\ell$$

Summarizing this reasoning in the form of a table,

	Initial Conc. (mol/ℓ)	**Change**	**Equil. Conc. (mol/ℓ)**
H_2	0.00	$+x$	x
I_2	0.00	$+x$	x
HI	1.20	$-2x$	$1.20 - 2x$

The algebraic equation must then be:

$$\frac{x^2}{(1.20 - 2x)^2} = 0.010$$

You may recall that we solved this equation in Example 8.2 by extracting the square root of both sides. There we found two values of x, 0.10 and −0.15. The negative value is chemically absurd; we cannot have a *negative* concentration. We conclude that in this chemical system, x = 0.10. Hence,

$$[H_2] = 0.10\ \text{mol}/\ell; \qquad [I_2] = 0.10\ \text{mol}/\ell$$

$$[HI] = (1.20 - 0.20)\text{mol}/\ell = 1.00\ \text{mol}/\ell$$

Example 8.5 When acetic acid, $HC_2H_3O_2$, dissociates in water, the following equilibrium system is established:

$$HC_2H_3O_2\ (aq) \rightleftharpoons H^+\ (aq) + C_2H_3O_2^-\ (aq)$$

The equilibrium constant, K_a, is 1.80×10^{-5}. Suppose we start with a concentration of $HC_2H_3O_2$ of 0.100 mol/ℓ. Take the initial concentrations of H^+ and $C_2H_3O_2^-$ to be zero.

a. Set up an algebraic equation in one unknown relating the equilibrium concentrations of $HC_2H_3O_2$, H^+, and $C_2H_3O_2^-$.
b. Solve this equation to obtain the equilibrium concentrations.

Solution

a. Let us choose our unknown, x, to be the equilibrium concentration of H^+. Looking at the balanced equation, we see that for every mole of H^+ formed, one mole of $C_2H_3O_2^-$ is formed and one mole of $HC_2H_3O_2$ is consumed. Hence, if we form x mol/ℓ of H^+, we must form x mol/ℓ of $C_2H_3O_2^-$ and use up x mol/ℓ of $HC_2H_3O_2$. Since we started with no $C_2H_3O_2^-$, its concentration at equilibrium, like that of H^+, must be x. Starting with 0.100 mol/ℓ of $HC_2H_3O_2$ and consuming x mol/ℓ leaves us with $(0.100 - x)$ mol/ℓ at equilibrium. In tabular form,

	Initial Conc. (mol/ℓ)	**Change**	**Equil. Conc. (mol/ℓ)**
$HC_2H_3O_2$	0.100	$-x$	$0.100 - x$
H^+	0.000	$+x$	x
$C_2H_3O_2^-$	0.000	$+x$	x

The equilibrium constant expression has the form

$$K_a = 1.80 \times 10^{-5} = \frac{[H^+] \times [C_2H_3O_2^-]}{[HC_2H_3O_2]}$$

Hence our algebraic equation becomes

$$1.80 \times 10^{-5} = \frac{(x)(x)}{0.100 - x} = \frac{x^2}{0.100 - x}$$

b. This equation was solved in Example 8.3, where we found that $x = 1.33 \times 10^{-3}$ or -1.35×10^{-3}. Here, as in Example 8.4, the negative root is absurd. We cannot have a concentration of H^+ less than zero. We conclude that

$$x = 1.33 \times 10^{-3}\ \text{mol/ℓ}$$

Hence $[H^+] = [C_2H_3O_2^-] = 1.33 \times 10^{-3}$ mol/ℓ

$$[HC_2H_3O_2] = 0.100\ \text{mol/ℓ} - 0.00133\ \text{mol/ℓ}$$

$$= 0.099\ \text{mol/ℓ}$$

Example 8.6 Consider the equilibrium

$$2CO_2\ (g) \rightleftharpoons 2\ CO\ (g) + O_2\ (g); \qquad K_c = 5.0 \times 10^{-4}$$

Suppose we start with pure CO_2 at a concentration of 1.00 mol/ℓ. No CO or O_2 are present initially. Set up an algebraic equation relating the equilibrium concentrations of CO, CO_2, and O_2.

Solution

It is ordinarily simplest to choose our unknown, x, to be the change in the concentration of a species which has a coefficient of 1 in the chemical equation. In this system, that species is O_2. We choose

$$x = [O_2]$$

The chemical equation tells us that, in forming 1 mol of O_2, we form 2 mol of CO and consume 2 mol of CO_2. Hence, the concentration of CO must have increased by $2x$ mol/ℓ; that of CO_2 must have decreased by $2x$ mol/ℓ. Recall that we started with no CO and 1.00 mol/ℓ of CO_2. Hence

$$[CO] = 2x$$

$$[CO_2] = 1.00 - 2x$$

The expression for K_c is

$$K_c = 5.0 \times 10^{-4} = \frac{[CO]^2 \times [O_2]}{[CO_2]^2}$$

Substituting $$5.0 \times 10^{-4} = \frac{(2x)^2(x)}{(1.00 - 2x)^2} = \frac{4x^3}{(1.00 - 2x)^2}$$

This is a third degree equation which cannot be solved by any of the methods discussed to this point. Its solution will be considered in Section 8.4.

8.4 SOLUTION OF SECOND AND HIGHER DEGREE EQUATIONS BY APPROXIMATION METHODS

Although it is always possible to solve a second degree equation by means of the quadratic formula, that approach is tedious at best. Recall Example 8.3, where we solved the equation

$$\frac{x^2}{0.100 - x} = 1.80 \times 10^{-5}$$

A considerable amount of arithmetic was involved, to say the least. Equations similar to this arise frequently in general chemistry; the one just cited applied to the dissociation of acetic acid (Example 8.5). We would like to find a simpler way to solve such equations. One approach, which we will discuss here, is called the method of **successive approximations.** It is most useful in problems dealing with equilibria in solutions of weak acids and weak bases.

To illustrate how this method works, let us apply it to the equation

$$\frac{x^2}{0.100 - x} = 1.80 \times 10^{-5}$$

Referring to Example 8.5, you will recall that x is the concentration of H^+ in a 0.100 M acetic acid solution ($K_a = 1.80 \times 10^{-5}$). Note that the ionization constant of acetic acid, 1.80×10^{-5}, is a very small number. Hence it seems likely that x, the concentration of H^+ produced when acetic acid ionizes, will be very small. In particular, we expect *x to be very much smaller than 0.100,* the original concentration of acetic acid. If this is true, we can neglect the x in the denominator on the left side of the above equation, writing

$$\frac{x^2}{0.100} = 1.80 \times 10^{-5}$$

This equation is readily solved for x:

$$x^2 = 1.80 \times 10^{-6}; \qquad x = \sqrt[2]{1.80 \times 10^{-6}} = 1.34 \times 10^{-3}$$

Compare this value of x, obtained by making the approximation $0.100 - x \approx 0.100$, to that obtained in Example 8.3, where we used the quadratic formula. The two numbers, 1.34×10^{-3} and 1.33×10^{-3}, differ from each other by 1 part in 133, or less than 1 percent. Ordinarily, errors as small as this are acceptable in working equilibrium problems. Equilibrium constants themselves are seldom valid to better than ± 5 percent.

We conclude that the assumption described above is valid for this equation. The question arises as to the general validity of this approach. If we are dealing with an equation of the type

$$\frac{x^2}{a - x} = K \tag{8.5}$$

under what conditions can we ignore the x in the denominator, so that

$$x^2 \approx aK; \qquad x \approx (aK)^{1/2} \quad ?$$

In general, we will regard this approximation as valid, provided the value of x obtained is no more than 5% of a. That is,

$$a - x \approx a \quad \text{if} \quad x \leq 5\%\, a \tag{8.6}$$

In most cases dealing with equilibria involving weak acids or bases, this condition is met and the approximation is valid. In the case of 0.100 M acetic acid, the value of x calculated by making the approximation is only a little more than 1% of the original concentration:

$$\frac{x}{a} = \frac{1.34 \times 10^{-3}}{1.00 \times 10^{-1}} = 0.0134 = 1.34\%$$

Sometimes, though, the error arising from the approximation will exceed the 5% limit we have set. If this happens, all is not lost. We can refine our calculation by making a second approximation, more nearly valid than the first. What we do here is to substitute for x, in the denominator of Equation 8.5, the value obtained by the first approximation. Solving the resulting equation for x gives a number much closer to the true value. The procedure followed is shown in Example 8.7.

Example 8.7 The weak acid HSO_4^- has an ionization constant of 1.0×10^{-2}:

$$HSO_4^- \text{ (aq)} \rightleftharpoons H^+ \text{ (aq)} + SO_4^{2-} \text{ (aq)}; \qquad K_a = 1.0 \times 10^{-2} = \frac{[H^+] \times [SO_4^{2-}]}{[HSO_4^-]}$$

Calculate $[H^+]$ in a solution prepared by adding one mole of HSO_4^- to one liter of water, making the initial concentration of HSO_4^- 1.00 M.

Solution

Following the same line of reasoning as in Example 8.5, we arrive at the following table:

	Initial Conc. (mol/ℓ)	**Change**	**Equil. Conc. (mol/ℓ)**
HSO_4^-	1.00	$-x$	$1.00 - x$
H^+	0.00	$+x$	x
SO_4^{2-}	0.00	$+x$	x

This leads to the algebraic equation

$$\frac{x^2}{1.00 - x} = 1.0 \times 10^{-2}$$

Making the approximation: $1.00 - x \approx 1.00$, we obtain

$$x^2 \approx 1.0 \times 10^{-2} \qquad x \approx 1.0 \times 10^{-1} = 0.10$$

In this case, the error exceeds that allowed under the "5% rule." That is,

$$\frac{x}{a} = \frac{0.10}{1.00} = 10\%$$

To obtain a better value for x, we substitute $x = 0.10$ in the denominator of the original equation. This gives us

$$\frac{x^2}{1.00 - 0.10} = \frac{x^2}{0.90} = 1.0 \times 10^{-2}$$

$$x^2 = 0.90 \times 10^{-2}; \qquad x = 0.95 \times 10^{-1} = 0.095 \text{ M} = [H^+]$$

This value of $[H^+]$, 0.095 M, is closer to the true concentration than that obtained from our first approximation, 0.10 M. This must be true, since 0.90 is a better approximation to the equilibrium concentration of HSO_4^- than was our first guess, 1.0. If we still are not satisfied, we can go one step further. Using the value of $[H^+]$ just calculated, we can attempt to obtain an even better value for $[HSO_4^-]$. If we do this, we find that $[HSO_4^-]$ stays at 0.90. That is,

$$[HSO_4^-] = 1.00 - 0.095 = 0.90 \text{ M} \qquad \text{(two significant figures)}$$

This means that if we were to solve again for $[H^+]$, we would get the same answer. In other words, the value 0.095 M is as close as we are going to get to the concentration of H^+. It is indeed the value we would get by applying the quadratic formula to the original second degree equation. (Try it!)

This technique can be applied to a wide variety of problems dealing with equilibria involving weak acids or bases. Usually, the first approximation: $a - x \approx a$, will be sufficient to obtain an answer that meets the 5% rule. Sometimes, as in Example 8.7, a second approximation will be necessary. Almost never do we need to go beyond this point.

The approximation method we have described is by no means restricted to second degree equations. Indeed, it is particularly useful for higher degree equations where exact solutions are very difficult or impossible to obtain. Consider, for example, the equation obtained in Example 8.6:

$$\frac{4x^3}{(1 - 2x)^2} = 5.0 \times 10^{-4}$$

Noting that the equilibrium constant, 5.0×10^{-4}, is small, we might assume that

$$(1 - 2x) \approx 1$$

The equation would then become

$$4x^3 \approx 5.0 \times 10^{-4}$$

$$x^3 \approx 1.25 \times 10^{-4}$$

$$x \approx 5.0 \times 10^{-2} = 0.050$$

If we wish to refine our answer, we make a second approximation, substituting this value of x in the denominator of the original equation:

$$\frac{4x^3}{(1 - 0.10)^2} = \frac{4x^3}{(0.90)^2} \approx 5.0 \times 10^{-4}$$

$$4x^3 \approx (0.81)(5.0 \times 10^{-4}) = 4.05 \times 10^{-4}$$

$$x^3 \approx 1.01 \times 10^{-4}$$

$$x \approx 4.7 \times 10^{-2} = 0.047$$

You can readily demonstrate that if a third approximation is made, the value of x remains unchanged.

A modification of the approximation method just described can be applied to solve any algebraic equation. This completely general approach is often referred to as the "guess and go" method. To use it, you first make an educated guess as to a value of x that will satisfy the equation. You then check to see how close you came to a true solution. Based on that check, you make a second, more accurate guess for x. This process is continued until you obtain a value of x that is accurate enough for your purposes.

To illustrate this approach, let us return to Example 8.6, with one change. Suppose the equilibrium constant is 0.45 instead of 5.0×10^{-4}. The algebraic equation now becomes

$$\frac{4x^3}{(1 - 2x)^2} = 0.45 \qquad (8.7)$$

Since the right side of Equation 8.7 is rather large, 0.45, it would *not* be reasonable to assume that $1 - 2x \approx 1$. However, all is not lost. You can make an educated guess as to the value of x by realizing that

1. x must be larger than zero. (Remember that x represents the concentration of O_2, which could hardly be negative!)
2. x must be smaller than 0.50. (Remember that you started with 1.00 mol of CO_2 and consumed $2x$ moles. If x were larger than 0.50, the equilibrium concentration of CO_2 would be negative.)

In summary, common sense tells us that in Equation 8.7

$$0.00 < x < 0.50$$

You really don't know, at this point, where x lies within this range. A reasonable guess might be the midpoint, 0.25. So, assume $x = 0.25$ and evaluate the left side of Equation 8.7:

$$\text{if } x = 0.25, \qquad \frac{4x^3}{(1 - 2x)^2} = \frac{4(0.25)^3}{(1 - 0.50)^2} = 0.25$$

The value just calculated, 0.25, is smaller than that required by Equation 8.7, 0.45. So, your guess of $x = 0.25$ must have been too small. Try a somewhat larger value, $x = 0.30$:

$$\text{if } x = 0.30, \qquad \frac{4x^3}{(1 - 2x)^2} = \frac{4(0.30)^3}{(1 - 0.60)^2} = 0.68$$

This time the answer is too large ($0.68 > 0.45$). Apparently the second guess, 0.30, was too large. From the two calculations, it appears that x must lie between 0.25 (too small) and 0.30 (too large). You could continue to refine your estimate of x, narrowing in on the correct value. This is done in Table 8.2; clearly, x is about 0.28.

Table 8.2 SOLUTION OF EQUATION 8.7 BY THE "GUESS AND GO" METHOD

Guess for x	*Value of* $\frac{4x^3}{(1 - 2x)^2}$	*Comparison*	*Conclusion*
1. $x = 0.25$	0.25	$0.25 < 0.45$	$x > 0.25$
2. $x = 0.30$	0.68	$0.68 > 0.45$	$0.30 > x > 0.25$
3. $x = 0.27$	0.37	$0.37 < 0.45$	$0.30 > x > 0.27$
4. $x = 0.28$	0.45	$0.45 = 0.45$	$x = 0.28$

To solve this equation, it took four guesses to "zero in" on the value of x. It seldom takes more than that, provided your first guess is in the right ballpark. With your calculator, you can obtain the numbers listed in Table 8.2 in a few minutes. With a little experience, you will find that the "guess and go" method is very effective in solving a wide variety of otherwise insoluble equations.

PROBLEMS

8.1 To make an angel cake, several ingredients are required, including 1 cup of flour and $\frac{3}{2}$ cup of sugar. We might say that

$$1 \text{ cup flour} + \tfrac{3}{2} \text{ cup sugar} \rightarrow 1 \text{ cake}$$

a. How many cups of flour are used to make 2 cakes? how many cups of sugar?

b. How many cups of flour are used to make *n* cakes? how many cups of sugar?

c. If you start with 7 cups of sugar, how many are left after making 2 cakes? *n* cakes?

8.2 A widget is made by putting 2 nuts on a bolt. In other words,

$$1 \text{ bolt} + 2 \text{ nuts} \rightarrow 1 \text{ widget}$$

Complete the following table:

	Orig. Amt.	Change	Final Amt.
Bolts	52	______	______
Nuts	94	______	______
Widgets	12	$+x$	______

8.3 Glass is made by melting together soda ash, limestone, and sand in a mass ratio of about 1 : 1 : 4. We might say, for example, that

$$1 \text{ kg soda ash} + 1 \text{ kg limestone} + 4 \text{ kg sand} \rightarrow 6 \text{ kg glass}$$

Complete the following tables:

	Orig. Amt.	Change	Final Amt.
soda ash	90 kg	______	______
limestone	70 kg	______	______
sand	500 kg	______	______
glass	0 kg	______	60 kg

	Orig. Amt.	Change	Final Amt.
soda ash	90 kg	______	______
limestone	70 kg	______	______
sand	500 kg	______	______
glass	0 kg	______	x kg

8.11 Consider the equilibrium

$$2\ HI\ (g) \rightleftharpoons H_2\ (g) + I_2\ (g)$$

Suppose the original concentration of HI is 0.40 mol/ℓ, while those of H_2 and I_2 are both zero. Take the equilibrium concentration of H_2 to be x. Express, in terms of x,

a. $[I_2]$ *b*. [HI]

8.12 Consider the equilibrium

$$HF\ (aq) \rightleftharpoons H^+\ (aq) + F^-\ (aq)$$

We start with pure HF, taking the original concentrations of H^+ and F^- to be zero. If we let the equilibrium concentration of H^+ be x, what will be

a. $[F^-]$?

b. [HF] if its original concentration is 1.00 M?

c. [HF] if its original concentration is 0.100 M?

8.13 Consider the equilibrium

$$N_2\ (g) + 3\ H_2\ (g) \rightleftharpoons 2\ NH_3\ (g)$$

Complete the following tables:

	Orig. Conc.	Change	Equil. Conc.
N_2	2.04	______	______
H_2	4.62	______	______
NH_3	0.00	______	1.00

	Orig. Conc.	Change	Equil. Conc.
N_2	2.04	$-x$	______
H_2	4.62	______	______
NH_3	0.00	______	______

8.4 Solve the following quadratic equations by reducing to first degree:

a. $x^2 = 2.0 \times 10^{-5}$

b. $\dfrac{x^2}{(1 - x)^2} = 6.0$

c. $\dfrac{x^2}{(0.10 - 2x)^2} = 1.0$

8.5 Solve the following equations for x:

a. $\dfrac{x^3}{(1 - x)^3} = 18$

b. $\dfrac{(2 - x)^2}{(2 + 2x)^2} = 16$

c. $(1 - x)^2 = 0.064x^2$

8.6 Use the quadratic formula to find x.

a. $\dfrac{x^2}{1.00 - x} = 1.00 \times 10^{-1}$

b. $\dfrac{x^2}{1.00 - x} = 1.00 \times 10^{-2}$

c. $\dfrac{x^2}{1.00 - x} = 1.00 \times 10^{-3}$

8.7 Find x, using the quadratic formula.

a. $\dfrac{4x^2}{2 - x} = 20$

b. $\dfrac{3 - x}{x^2} = 0.10$

c. $\dfrac{x(0.50 + x)}{1.00 - x} = 0.010$

8.8 Find x, using the quadratic formula.

a. $\dfrac{x^2}{1.0 - x} = 0.30$

b. $\dfrac{x}{(1.00 - 2x)^2} = 5.00$

8.14 For the equilibrium referred to in Problem 8.11, take $K_c = 0.015$.

a. Set up an expression for K_c in terms of $[H_2]$, $[I_2]$, and $[HI]$.

b. Taking $[H_2]$, $[I_2]$, and $[HI]$ to be x, x, and $0.40 - 2x$, in that order, determine x.

8.15 For the equilibrium referred to in Problem 8.11,

a. Complete the following table:

	Orig. Conc.	Change	Equil. Conc.
H_2	1.00	$-x$	______
I_2	1.00	______	______
HI	0.00	______	______

b. Set up an algebraic equation and solve it for x, taking $K_c = 0.015$.

8.16 Consider the equilibrium in Problem 8.12, for which $K_a = 7.00 \times 10^{-4}$.

a. Set up an expression for K_a in terms of $[H^+]$, $[F^-]$, and $[HF]$.

b. Taking $[H^+]$, $[F^-]$, and $[HF]$ to be x, x, and $1.00 - x$, in that order, solve for x using the quadratic formula.

8.17 For the equilibrium:

$$2\ HI\ (g) \rightleftharpoons H_2\ (g) + I_2\ (g);\ K_c = 0.015$$

a. Complete the following table:

	Orig. Conc.	Change	Equil. Conc.
H_2	0.00	$+x$	______
I_2	0.50	______	______
HI	1.00	______	______

b. Set up an algebraic equation and solve it for x, using the quadratic formula.

8.18 Consider the equilibrium

$$PCl_5\ (g) \rightleftharpoons PCl_3\ (g) + Cl_2\ (g);\ K_c = 0.20$$

Suppose you start with a concentration of PCl_5 of 1.00 mol/ℓ, with no PCl_3 or Cl_2. Set up an algebraic equation in x and solve it, using the quadratic formula, to obtain the equilibrium concentrations of Cl_2, PCl_3, and PCl_5.

8.9 Consider the equations given in Problem 8.6. In each case, solve by neglecting the x in the denominator, taking $1.00 - x = 1.00$. What is the percent difference in each case from the answer obtained by using the quadratic formula?

8.10 Solve the equation

$$\frac{x^2}{1.00 - x} = 2.0 \times 10^{-2}$$

by the method of successive approximations. (Consider only the positive root.)

8.19 Repeat Problem 8.16, but this time solve by neglecting the x in the $1.00 - x$ term. What is the percent difference between the answer obtained this way and that found using the quadratic formula? Is the answer obtained here acceptable in terms of the 5% rule?

8.20 Suppose that in Example 8.7 the solution were prepared by adding two moles of HSO_4^- to one liter of water. Calculate $[H^+]$ by the method of successive approximations.

CHAPTER

9 Functional Relationships

In chemistry and the other natural sciences, we often deal with functional relationships between two or more variables. An example of such a relationship is that between the volume, V, of a sphere and its radius, r:

$$V = \frac{4}{3}\pi r^3 \tag{9.1}$$

Here, the two "variables" are V and r. Depending upon the size of the sphere, the radius, r, might have any value, such as 1.00 cm, 2.00 cm, . . . Once we settle upon a value of r, we can use Equation 9.1 to calculate the corresponding value of the other variable, V. We find, for example, that $V = 4.19\ \text{cm}^3$ if $r = 1.00$ cm; $V = 33.5\ \text{cm}^3$ if $r = 2.00$ cm, and so on.

In general, a variable (y) is said to be a function of another variable (x) if, for various values of x, we can calculate corresponding values of y. We describe this situation by writing the equation

$$y = f(x)$$

The variable x to which we first assign a numerical value is called the **independent variable.** The other variable, y, whose value depends upon that assigned to x, is called the **dependent variable.** In Equation 9.1, we would refer to r as the independent variable and V as the dependent variable.

The functional relationship between two variables may be expressed in any of three different ways:

(1) An *algebraic* equation which allows us to calculate y for any value of x. A simple example is

$$y = 2x$$

This equation tells us that y can be found by multiplying x by 2.

(2) A *table* which lists values of y corresponding to selected values of x. Table 9.1 does this for the function $y = 2x$.

Table 9.1

y	0	2	4	6	8	10
x	0	1	2	3	4	5

(3) A *graph* drawn through points indicating corresponding values of y and x. The graph of the function $y = 2x$ between $x = 0$ and $x = 5$ is shown in Figure 9.1. The points shown are those listed in Table 9.1. The line drawn through the points allows us to find values of y corresponding to x values not listed in the table.

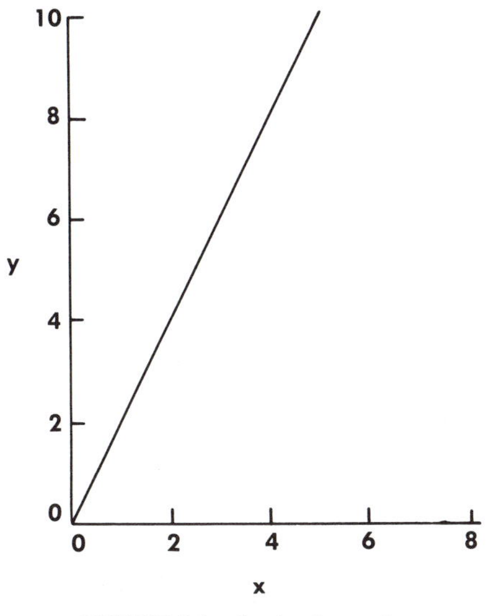

FIGURE 9.1 Graph of $y = 2x$.

In this chapter, we shall see how several different types of functional relationships can be expressed by means of equations and tables. Chapter 10 is devoted to ways of representing functional relationships by means of graphs. Throughout both chapters, the examples we deal with will be taken from general chemistry.

9.1 DIRECT PROPORTIONALITY

$$y = ax; \quad \frac{y_2}{y_1} = \frac{x_2}{x_1}$$

Many functional relationships in general chemistry fall in this category. Listed in Table 9.2 are two examples in which one variable is directly proportional to another. Looking at these two sets of data, we see that they have a common property. In both cases, the quotient y/x has a constant value throughout (0.082 in the first case, 0.030 in the second).

Table 9.2 EXAMPLES OF DIRECT PROPORTIONALITY IN GENERAL CHEMISTRY

VOLUME, V, OF 1 MOL OF AN IDEAL GAS AT 1 ATM AS A FUNCTION OF ABSOLUTE TEMPERATURE, T

V (ℓ)	8.2	16.4	24.6	32.8	41.0
T (K)	100	200	300	400	500
V/T (ℓ/K)	0.082	0.082	0.082	0.082	0.082

RATE OF DECOMPOSITION OF N_2O_5 AT 45°C AS A FUNCTION OF CONCENTRATION, M

Rate (M/min)	0.0030	0.0060	0.0090	0.0120	0.0150
Conc. (M)	0.10	0.20	0.30	0.40	0.50
Rate/Conc. (min^{-1})	0.030	0.030	0.030	0.030	0.030

Algebraically, we can say that y is directly proportional to x if, for all values of the two variables,

$$\frac{y}{x} = a \quad \text{or} \quad y = ax \tag{9.2}$$

Here, a is a "constant," often called a proportionality constant. It retains the same value throughout, regardless of the individual values of x and y. In the direct proportionalities listed in Table 9.2, a is 0.082 ℓ/K in one case and 0.030/min in the other. The relation given in Table 9.1, $y = 2x$, is also a direct proportionality, with $a = 2$.

The defining equation for direct proportionality, Equation 9.2, leads to a simple "two-point" equation relating "final" and "initial" values of y and x. Using the subscripts 2 and 1 to represent final and initial states,

$$y_2 = ax_2$$

$$y_1 = ax_1$$

Dividing the first equation by the second, the constant a drops out, and we have

$$\frac{y_2}{y_1} = \frac{x_2}{x_1} \tag{9.3}$$

Example 9.1 The solubility of a gas in a liquid is directly proportional to its partial pressure. The solubility of O_2 in water at 25°C is 0.085 mol/ℓ at one atmosphere. Calculate its solubility at a partial pressure of 0.20 atm.

Solution

Using the letters S and P to represent solubility and pressure in that order, we have (from Equation 9.3)

$$\frac{S_2}{S_1} = \frac{P_2}{P_1}$$

Taking P_1 to be 1.00 atm and P_2 to be 0.20 atm, S_1 is 0.085 mol/ℓ, and we have

$$\frac{S_2}{0.085 \text{ mol}/\ell} = \frac{0.20 \text{ atm}}{1.00 \text{ atm}} = 0.20$$

$$S_2 = 0.20(0.085 \text{ mol}/\ell) = 0.017 \text{ mol}/\ell$$

Sometimes we find that y is directly proportional, not to x itself, but to some positive power of x. That is,

$$y = ax^n; \quad n > 0 \tag{9.4}$$

Examples of such relationships in chemistry include

— the relation between the velocity of a gas molecule, u, and its absolute temperature, T:

$$u = aT^{1/2}; \quad n = \tfrac{1}{2}$$

— the relation between the translational kinetic energy, E, of a gas molecule and its velocity, u:

$$E = au^2; \quad n = 2$$

We can easily obtain the two-point equation corresponding to Equation 9.4.

Again, the subscripts 2 and 1 represent final and initial states, in that order.

$$y_2 = ax_2^n$$

$$y_1 = ax_1^n$$

We divide the first equation by the second; the constant a cancels and we get

$$\frac{y_2}{y_1} = \frac{x_2^n}{x_1^n} = \left(\frac{x_2}{x_1}\right)^n \qquad (9.5)$$

Equation 9.5 is particularly useful in the area of chemical kinetics. Specifically, it is used to determine what is known as the *order* of a reaction from rate data for that reaction. For a single reactant, A, the general expression for the relationship between rate and concentration is

$$\text{rate} = \text{k}(\text{conc. } A)^n \qquad (9.6)$$

Here, k is called the rate constant and the exponent n is known as the order of the reaction. If $n = 1$, the reaction is said to be first order in A; if $n = 2$, the reaction is second order in A, and so on. The value of n can be determined by measuring rate as a function of concentration and applying Equation 9.5 (Example 9.2).

Example 9.2 Nitrogen dioxide, NO_2, decomposes as follows:

$$2\ NO_2(g) \rightarrow 2\ NO(g) + O_2(g)$$

The rate of this reaction was measured at a series of different concentrations with the following results:

conc. NO_2 (M)	0.10	0.20	0.30	0.40
rate (M/s)	0.020	0.080	0.180	0.320

Using these data with Equation 9.5, find the order of the reaction with respect to nitrogen dioxide. That is, find the value of n in the equation

$$\text{rate} = \text{k}(\text{conc. } NO_2)^n$$

Solution

Applying Equation 9.5, we have

$$\frac{\text{rate}_2}{\text{rate}_1} = \left(\frac{\text{conc.}_2}{\text{conc.}_1}\right)^n$$

Let us choose our two concentrations to be 0.20 and 0.10 M:

$$\frac{0.080}{0.020} = \left(\frac{0.20}{0.10}\right)^n$$

Simplifying: $4 = 2^n$

Clearly, n is 2. The same result would be obtained if we chose any other pair of data points.

Once the value of n in Equation 9.6 is known, we can readily calculate how the reaction rate will change with concentration (Example 9.3).

Example 9.3 Given that the rate expression for the decomposition of nitrogen dioxide is

$$\text{rate} = k(\text{conc. } NO_2)^2$$

and that the rate is 0.020 M/s when the concentration of NO_2 is 0.10 M, calculate the rate when the concentration is 1.0 M.

Solution

We know that n in Equation 9.5 must be 2. Letting y represent rate and x represent concentration, we have

$$\frac{\text{rate}_2}{\text{rate}_1} = \left(\frac{\text{conc.}_2}{\text{conc.}_1}\right)^2$$

We take "rate$_2$" to be the rate when the concentration is 1.0 M; "rate$_1$" corresponds to a concentration of 0.10 M.

$$\text{rate}_2 = \text{rate}_1 \times \left(\frac{\text{conc.}_2}{\text{conc.}_1}\right)^2$$

$$= 0.020\,\frac{\text{M}}{\text{s}} \times \left(\frac{1.0}{0.10}\right)^2 = 0.020\,\frac{\text{M}}{\text{s}} \times 100 = 2.0\,\frac{\text{M}}{\text{s}}$$

9.2 INVERSE PROPORTIONALITY

$$y = \frac{a}{x}; \quad \frac{y_2}{y_1} = \frac{x_1}{x_2}$$

A quantity y is said to be **inversely proportional** to x if their product, yx, is the same for all values of x:

$$yx = a; \quad \text{or} \quad y = \frac{a}{x} \tag{9.7}$$

Here, as in Equations 9.2 and 9.4, a is a constant, independent of x or y. Two examples of inverse proportionality in general chemistry are shown in Table 9.3. Note that in both cases the product of the two variables is constant throughout.

Table 9.3 EXAMPLES OF INVERSE PROPORTIONALITY IN GENERAL CHEMISTRY

VOLUME, V, OF 1 MOL OF AN IDEAL GAS AT 25°C AS A FUNCTION OF PRESSURE, P					
V (ℓ)	24.5	12.2	8.16	6.12	4.89
P (atm)	1.00	2.00	3.00	4.00	5.00
$P \times V$ (ℓ · atm)	24.5	24.4	24.5	24.5	24.4
CONC. OF H^+ IN AQUEOUS SOLUTION AT 25°C AS A FUNCTION OF CONC. OH^-					
Conc. H^+ (M)	1.0×10^{-14}	1.0×10^{-10}	1.0×10^{-7}	1.0×10^{-4}	
Conc. OH^- (M)	1.0	1.0×10^{-4}	1.0×10^{-7}	1.0×10^{-10}	
Conc. H^+ × Conc. OH^-	1.0×10^{-14}	1.0×10^{-14}	1.0×10^{-14}	1.0×10^{-14}	

Example 9.4 Use the information in Table 9.3 to calculate the concentration of OH^- in a solution in which the concentration of H^+ is 5.0×10^{-2} M.

Solution

The basic relation is

$$\text{conc. } H^+ \times \text{conc. } OH^- = 1.0 \times 10^{-14}$$

Solving for conc. OH^-

$$\text{conc. } OH^- = \frac{1.0 \times 10^{-14}}{\text{conc. } H^+} = \frac{1.0 \times 10^{-14}}{5.0 \times 10^{-2}} = 0.20 \times 10^{-12} = 2.0 \times 10^{-13} \text{ M}$$

The two-point equation for an inverse proportionality is readily obtained. Writing Equation 9.7 for both final and initial states,

$$y_2x_2 = a; \quad y_1x_1 = a$$

Since y_2x_2 and y_1x_1 are both equal to a, they must be equal to each other:

$$y_2x_2 = y_1x_1$$

or

$$\frac{y_2}{y_1} = \frac{x_1}{x_2} \tag{9.8}$$

In some cases, y is inversely proportional, not to x itself, but to some positive power of x. That is,

$$yx^n = a \quad \text{or} \quad y = \frac{a}{x^n} \tag{9.9}$$

An example of such a relationship in general chemistry is Graham's law of effusion. We find that the rate of effusion of a gas is inversely proportional to the square root of its molecular mass, MM:

$$\text{rate} \times (\text{MM})^{1/2} = \text{constant}$$

or

$$\frac{\text{rate}_2}{\text{rate}_1} = \frac{(\text{MM}_1)^{1/2}}{(\text{MM}_2)^{1/2}} = \left(\frac{\text{MM}_1}{\text{MM}_2}\right)^{1/2} \tag{9.10}$$

where the subscripts 2 and 1 refer to two gases of different molecular mass.

Example 9.5 Oxygen, O_2, has a molecular mass of 32.0. It effuses 2.41 times as fast as a certain unknown gas. What is the molecular mass of the unknown gas?

Solution

Using the subscript 2 to represent O_2 and 1 for the unknown gas,

$$\frac{\text{rate}_2}{\text{rate}_1} = 2.41 = \left(\frac{\text{MM}_1}{32.0}\right)^{1/2}$$

To solve, we square both sides of the equation:

$$(2.41)^2 = \frac{\text{MM}_1}{32.0}$$

$$\text{MM}_1 = 32.0\ (2.41)^2 = 32.0 \times 5.81 = 186$$

9.3 LINEAR FUNCTIONS

$$y = ax + b; \quad \frac{(y_2 - y_1)}{(x_2 - x_1)} = a$$

A **linear function** is one which has the general form

$$y = ax + b \tag{9.11}$$

where a and b are constants.* The phrase "linear function" is used because a straight line is obtained when y is plotted against x (see Chapter 10). The relation $y = ax$, discussed in Section 9.1, is a special case of a linear function with $b = 0$.

A simple example of a linear relationship is that between temperatures expressed in degrees Fahrenheit (°F) vs degrees Celsius (°C):

$$°F = 1.8(°C) + 32° \tag{9.12}$$

Comparing this equation with Equation 9.11, we see that in this case

$$a = 1.8; \quad b = 32$$

Referring to Figure 9.2, we see that °F is indeed a linear (straight-line) function of °C.

The most useful two-point equation for a linear function is obtained by writing Equation 9.11 for the two states:

$$y_2 = ax_2 + b$$

$$y_1 = ax_1 + b$$

and subtracting to eliminate b:

$$y_2 - y_1 = a(x_2 - x_1) \quad \text{or} \quad \frac{y_2 - y_1}{x_2 - x_1} = a \tag{9.13}$$

Equation 9.13 offers a simple way to check a set of data to see if it represents a linear function (Table 9.4). It is also useful in some problems in chemistry (see Example 9.6).

Table 9.4 TEST FOR A LINEAR FUNCTION: $(y_2 - y_1)/(x_2 - x_1)$ = CONSTANT

FUNCTION 1			*FUNCTION 2*			*FUNCTION 3*		
y	x	$\frac{(y_2 - y_1)}{(x_2 - x_1)}$	y	x	$\frac{(y_2 - y_1)}{(x_2 - x_1)}$	y	x	$\frac{(y_2 - y_1)}{(x_2 - x_1)}$
3	0		1	0		0	0	
		2/1 = 2			3/1 = 3			3/1 = 3
5	1		4	1		3	1	
		2/1 = 2			5/1 = 5			3/1 = 3
7	2		9	2		6	2	
		2/1 = 2			7/1 = 7			3/1 = 3
9	3		16	3		9	3	
linear			*non-linear*			*linear*		

*As we will see in Chapter 10, a is the slope of the straight line obtained by plotting y against x. The constant b is the "y intercept," that is, the value of y when $x = 0$.

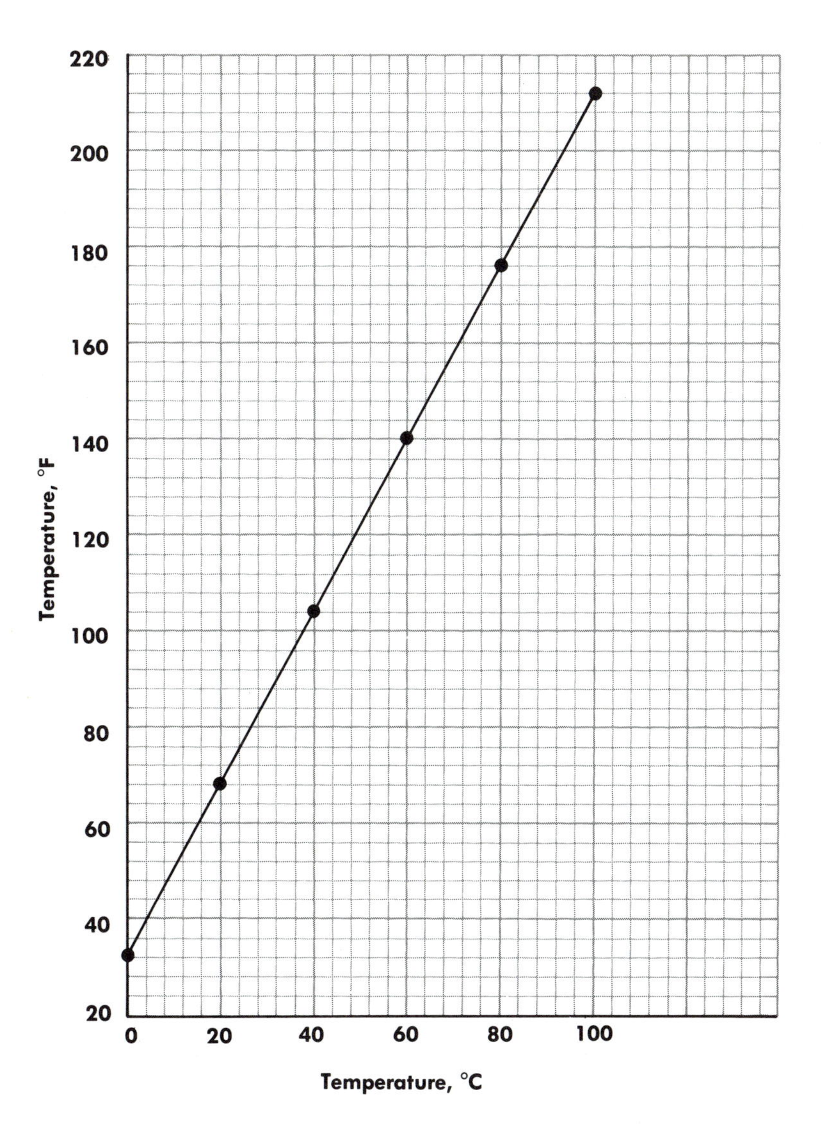

FIGURE 9.2 Graph of °F vs. °C from 0 to 100°C.

Example 9.6 Consider Equation 9.12, relating degrees Fahrenheit to degrees Celsius.

a. Write a two-point equation of the form of Equation 9.13 relating $°F_2$, $°F_1$, $°C_2$, and $°C_1$.
b. If the Celsius temperature increases by 15°, what is the corresponding increase in the Fahrenheit temperature?
c. A 20° increase in °F corresponds to what change in °C?

Solution

a. Here, a is 1.8, so Equation 9.13 becomes

$$\frac{°F_2 - °F_1}{°C_2 - °C_1} = 1.8$$

b. If the Celsius temperature increases by 15°C, then $°C_2 - °C_1 = 15°$. Using the equation just obtained, we get

$$\frac{°F_2 - °F_1}{15°} = 1.8; \quad °F_2 - °F_1 = 1.8(15°) = 27°$$

We conclude that the Fahrenheit temperature increased by 27°.

c. Here, the increase in °F is 20°. Hence,

$$°C_2 - °C_1 = \frac{°F_2 - °F_1}{1.8} = \frac{20}{1.8} = 11°$$

A basic equation of thermodynamics, called the Gibbs-Helmholtz equation, is a linear function. It relates the *free energy change,* ΔG, of a reaction to the absolute temperature, T:

$$\Delta G = \Delta H - T\Delta S \tag{9.14}$$

To a good approximation, the quantities ΔH and ΔS in this equation are independent of temperature. Hence, they can be considered to be constants for a particular reaction. The two quantities ΔH and ΔS are referred to as the *enthalpy change* and the *entropy change,* in that order. Comparing Equation 9.14 to the basic equation for a linear function, $y = ax + b$, we see that

$$T = \text{independent variable } (x)$$

$$\Delta G = \text{dependent variable } (y)$$

$$\Delta H = \text{constant } (b)$$

$$-\Delta S = \text{constant } (a)$$

Example 9.7 Using Equation 9.14, determine

a. ΔG at 400 K for a reaction in which $\Delta H = 61.0$ kJ and $\Delta S = 0.020$ kJ/K.
b. ΔS for a reaction in which $\Delta H = -32.0$ kJ and $\Delta G = -20.0$ kJ at 300 K.
c. the temperature at which $\Delta G = 0$ for the reaction in (b).

Solution

a. $\Delta G = 61.0 \text{ kJ} - 400 \text{ K}(0.020 \text{ kJ/K}) = 53.0 \text{ kJ}$
b. Solving Equation 9.14 for ΔS,

$$\Delta S = \frac{\Delta H - \Delta G}{T} = \frac{-32.0 \text{ kJ} + 20.0 \text{ kJ}}{300 \text{ K}} = -0.0400 \text{ kJ/K}$$

c. When $\Delta G = 0$, we see from Equation 9.14 that

$$\Delta H = T\Delta S \quad \text{or} \quad T = \frac{\Delta H}{\Delta S}$$

Substituting numbers from (b),

$$T = \frac{-32.0 \text{ kJ}}{-0.0400 \text{ kJ/K}} = 800 \text{ K}$$

9.4 LOGARITHMIC FUNCTIONS

In many of the functional relationships of general chemistry, one of the variables appears as a logarithmic term. Perhaps the most common logarithmic function in chemistry takes the form

$$\log y = \frac{-A}{x} + B \qquad (9.15)$$

where x is the independent variable, y is the dependent variable, and A and B are positive numbers (constants). Three examples of functional relationships of this type are given in Table 9.5. Notice that

— in each case, the independent variable x is the temperature T, in K.

— the dependent variable y may be the vapor pressure of a liquid P, the equilibrium constant for a reaction K, or the rate constant k.

— the constant A in each case involves an energy term (heat of vaporization, heat of reaction, energy of activation) divided by 2.303 R. In the vapor pressure equation, for example

$$A = \frac{\Delta H_{vap}}{2.303\ R}$$

Table 9.5 EXAMPLES OF THE FUNCTION LOG $y = \frac{-A}{x} + B$

FUNCTION	*MEANING OF TERMS*
$\log P = \frac{-\Delta H}{2.303\ RT} + B$	P = vapor pressure of liquid
	T = absolute temperature (K)
	R = gas constant = 8.314 J/(mol · K)
	ΔH = heat of vaporization (joules per mole)
$\log K = \frac{-\Delta H}{2.303\ RT} + B$	K = equilibrium constant for reaction at absolute temperature T
	ΔH = enthalpy change for reaction (joules)
$\log k = \frac{-E_a}{2.303\ RT} + B$	k = rate constant at T
	E_a = activation energy (joules)

The two-point equation corresponding to Equation 9.15 is obtained in a manner similar to that used for a linear function. We start by writing the equation twice, once for a "final" state (subscript 2) and once for an "initial" state (subscript 1).

$$\log y_2 = \frac{-A}{x_2} + B$$

$$\log y_1 = \frac{-A}{x_1} + B$$

Subtracting to eliminate the constant B,

$$\log y_2 - \log y_1 = \frac{-A}{x_2} + \frac{A}{x_1}$$

To simplify this equation, we factor the right side:

$$\log y_2 - \log y_1 = A\left[\frac{1}{x_1} - \frac{1}{x_2}\right]$$

Recall (Chapter 5) that $\log y_2 - \log y_1 = \log (y_2/y_1)$:

$$\log \frac{y_2}{y_1} = A\left[\frac{1}{x_1} - \frac{1}{x_2}\right]$$

This equation is ordinarily written in a somewhat different form, obtained by putting the quantity in brackets over a common denominator, x_1x_2.

$$\frac{1}{x_1} - \frac{1}{x_2} = \frac{x_2}{x_1x_2} - \frac{x_1}{x_1x_2} = \frac{x_2 - x_1}{x_1x_2}$$

Hence,

$$\log \frac{y_2}{y_1} = A\left[\frac{x_2 - x_1}{x_1x_2}\right] \tag{9.16}$$

Example 9.8 Consider the two-point equation relating the activation energy, E_a, to the rate constants, k_2 and k_1, at two different temperatures, T_2 and T_1:

$$\log \frac{k_2}{k_1} = \frac{E_a(T_2 - T_1)}{(2.303)(8.314)T_2T_1}$$

Calculate

a. the rate constant, k_2, at 310 K if k_1 at 300 K is 0.25/min and $E_a = 4.2 \times 10^4$ J.
b. the activation energy if k doubles when the temperature increases from 300 K to 310 K.
c. the temperature at which k is twice its value at 300 K, if $E_a = 8.60 \times 10^4$ J.

Solution

a. We can readily evaluate the right side of the equation:

$$\log \frac{k_2}{k_1} = \frac{4.2 \times 10^4\,(310 - 300)}{(2.303)(8.314)(310)(300)} = 0.24$$

Using a calculator, you should find that the antilog of 0.24 is 1.7. So,

$$\frac{k_2}{k_1} = 1.7$$

Since $k_1 = 0.25$/min, $k_2 = 1.7 \times 0.25$/min $= 0.42$/min

b. Note that $k_2/k_1 = 2$. Hence, the equation becomes

$$\log 2 = 0.301 = \frac{E_a(310 - 300)}{(2.303)(8.314)(310)(300)}$$

Solving for E_a,

$$E_a = \frac{(0.301)(2.303)(8.314)(310)(300)}{10} = 5.4 \times 10^4 \text{ J}$$

c. Taking T_1 to be 300 K,

$$\log 2 = 0.301 = \frac{8.60 \times 10^4(T_2 - 300)}{(2.303)(8.314)(300)T_2}$$

To find T_2, it may be simplest to first solve for the quantity $(T_2 - 300)/T_2$:

$$\frac{T_2 - 300}{T_2} = \frac{(0.301)(2.303)(8.314)(300)}{8.60 \times 10^4} = 0.0201$$

Solving: $0.9799T_2 = 300$

$$T_2 = 306 \text{ K}$$

Two-point equations for other types of logarithmic functions can be obtained in a manner similar to that used to get Equation 9.16. The approach used is illustrated in Example 9.9.

Example 9.9 Consider a so-called "first order" reaction, where the rate is directly proportional to the concentration of reactant. For such a reaction, the following equation applies:

$$\log X = \frac{-kt}{2.303} + B$$

Here, X is the concentration of reactant, k is the rate constant, and t is the time. Obtain a two-point relation involving X_2, X_1, t_2, and t_1.

Solution

As usual, we write the equations for the initial state (subscript 1) and the final state (subscript 2):

$$\log X_1 = \frac{-kt_1}{2.303} + B$$

$$\log X_2 = \frac{-kt_2}{2.303} + B$$

Subtracting the second equation from the first,

$$\log X_1 - \log X_2 = \frac{-kt_1}{2.303} + \frac{kt_2}{2.303}$$

Condensing: $\log \frac{X_1}{X_2} = \frac{k(t_2 - t_1)}{2.303}$

This is the equation called for in the statement of the problem. You may be familiar with it in a somewhat different form. Suppose we take t_1 to be "time zero" and X_1 to be the original concentration of reactant, X_0. The difference $(t_2 - t_1)$ now becomes the elapsed time t since the reaction started. The concentration of reactant at time t is written as X. With these modifications, the concentration–time relation for a first order reaction becomes

$$\log \frac{X_0}{X} = \frac{kt}{2.303}$$

9.5 "CONSTANTS" INVOLVED IN FUNCTIONAL RELATIONSHIPS

Throughout general chemistry, many different kinds of "constants" are used in expressing functional relationships. There is the gas constant, R, which appears in the ideal gas law

$$PV = nRT$$

and in several thermodynamic relationships, such as

$$\Delta G^\circ = -RT \ln K$$

Again, we talk about the rate constant, k, in expressions such as

$$\text{rate} = k(\text{conc. } NO_2)^2$$

In another case, we use the dissociation constant, K_a, in describing the equilibrium between acetic acid and its ions in water solution:

$$HC_2H_3O_2(aq) \rightleftharpoons H^+(aq) + C_2H_3O_2^-(aq)$$

$$K_a = \frac{[H^+] \times [C_2H_3O_2^-]}{[HC_2H_3O_2]}$$

To use expressions such as these properly, you must understand how constants such as R, k, and K_a are determined and what they mean. The following general principles apply.

1. GIVEN AN EQUATION INVOLVING A SINGLE CONSTANT, THAT CONSTANT CAN BE DETERMINED BY MEASURING SIMULTANEOUSLY THE VALUES OF ALL THE VARIABLES IN THE EQUATION.

Knowing, for example, that one mole of an ideal gas occupies 22.4 ℓ at 0°C (273 K) at one atmosphere pressure, we can calculate the gas constant R:

$$R = \frac{PV}{nT} = \frac{(1.00 \text{ atm})(22.4\ell)}{(1.00 \text{ mol})(273 \text{ K})} = 0.0821 \frac{\ell \cdot \text{atm}}{\text{mol} \cdot \text{K}}$$

Example 9.10 Given the information in Example 9.2, determine the rate constant k in the rate equation

$$\text{rate} = k(\text{conc. } NO_2)^2$$

Solution

Solving the equation for k,

$$k = \frac{\text{rate}}{(\text{conc. } NO_2)^2}$$

Note from Example 9.2 that the rate is 0.020 M/s when conc. NO_2 is 0.10 M. So,

$$k = \frac{0.020 \text{ M/s}}{(0.10 \text{ M})^2} = 2.0/(\text{M} \cdot \text{s})$$

If an equation contains two constants, two sets of values of the variables are required to evaluate the constants (Example 9.11).

Example 9.11 Consider Equation 9.14,

$$\Delta G = \Delta H - T\Delta S$$

where ΔH and ΔS are taken to be constants. For a certain reaction, $\Delta G = -32.0$ kJ at 300 K and $\Delta G = -40.0$ kJ at 500 K. Use this information to determine

a. ΔS b. ΔH

Solution

a. We start by writing the equation twice, once at 300 K and once at 500 K:

$$-32.0 \text{ kJ} = \Delta H - (300 \text{ K}) \Delta S \quad (1)$$

$$-40.0 \text{ kJ} = \Delta H - (500 \text{ K}) \Delta S \quad (2)$$

To find ΔS, we eliminate ΔH by subtracting (2) from (1):

$$-32.0 \text{ kJ} - (-40.0 \text{ kJ}) = -(300 \text{ K}) \Delta S - (-500 \text{ K}) \Delta S$$

$$8.0 \text{ kJ} = (200 \text{ K}) \Delta S$$

Solving: $\Delta S = 8.0 \text{ kJ}/200 \text{ K} = 0.040 \text{ kJ/K}$

b. Knowing ΔS, we can calculate ΔH from either Equation (1) or (2):

(1) $-32.0 \text{ kJ} = \Delta H - 300(0.040)\text{kJ}$

$\Delta H = -32.0 \text{ kJ} + 12.0 \text{ kJ} = -20.0 \text{ kJ}$

(2) $-40.0 \text{ kJ} = \Delta H - 500(0.040)\text{kJ}$

$\Delta H = -40.0 \text{ kJ} + 20.0 \text{ kJ} = -20.0 \text{ kJ}$

The two answers for ΔH are, of course, the same.

Example 9.11 illustrates a general procedure that can be used to evaluate the two constants a and b, in any linear function:

$$y = ax + b$$

We must know two different sets of y and x values to obtain both a and b. To obtain a, we use Equation 9.13:

$$a = \frac{y_2 - y_1}{x_2 - x_1}$$

Knowing y_2, x_2, y_1 and x_1, we can obtain a. Once the value of a is established, b can be found by substituting into the original equation. Either the final or initial conditions can be used. That is,

$$b = y_2 - ax_2 \quad \text{or} \quad b = y_1 - ax_1$$

2. The term "constant" implies that the quantity has the same value regardless of the values of the variables in the equation.

To illustrate this point, consider the following data for the dissociation of acetic acid in solutions of three different concentrations.

	$[HC_2H_3O_2]$	$[H^+]$	$[C_2H_3O_2^-]$	$K_a = \frac{[H^+] \times [C_2H_3O_2^-]}{[HC_2H_3O_2]}$
I	1.0	1.8×10^{-4}	1.0×10^{-1}	$\frac{(1.8 \times 10^{-4})(1.0 \times 10^{-1})}{1.0} = 1.8 \times 10^{-5}$
II	1.0×10^{-1}	9.0×10^{-3}	2.0×10^{-4}	$\frac{(9.0 \times 10^{-3})(2.0 \times 10^{-4})}{1.0 \times 10^{-1}} = 1.8 \times 10^{-5}$
III	2.0	6.0×10^{-3}	6.0×10^{-3}	$\frac{(6.0 \times 10^{-3})(6.0 \times 10^{-3})}{2.0} = 1.8 \times 10^{-5}$

Clearly, K_a remains the same regardless of the values of $[HC_2H_3O_2]$, $[H^+]$, and $[C_2H_3O_2^-]$. This is a consequence of the fact that there is a functional relationship between these three variables. Only two of them can be chosen independently. The other variable is fixed by the requirement that the quotient

$$\frac{[H^+] \times [C_2H_3O_2^-]}{[HC_2H_3O_2]}$$

always equals 1.8×10^{-5}.

The magnitude of a constant may and often does depend upon other variables which are not specified in the equation. Both rate constants and equilibrium constants are temperature-dependent. For example, the equation

$$\frac{[H^+] \times [F^-]}{[HF]} = 7.0 \times 10^{-4}$$

for the equilibrium between HF and its ions applies only at one temperature, 25°C. At other temperatures, the equilibrium constant will have some value other than 7.0×10^{-4}. At 100°C, for example, K_a for HF is 2.5×10^{-4}.

3. Constants frequently have units, in which case their numerical value will depend upon the units in which they are expressed.

In this chapter, we have used two different values for the gas law constant R: 0.0821 $\ell \cdot$ atm/(mol $\cdot$ K) and 8.31 J/(mol $\cdot$ K). Both these values involve the same amount of energy per mole per kelvin. The conversion from one set of units to the other is easily made, using the relation

$$101.3 \text{ J} = 1 \ \ell \cdot \text{atm}$$

$$R = 0.0821 \frac{\ell \cdot \text{atm}}{\text{mol} \cdot \text{K}} \times \frac{101.3 \text{ J}}{1 \ \ell \cdot \text{atm}} = 8.31 \text{ J/(mol} \cdot \text{K)}$$

PROBLEMS

9.1 Given that y is directly proportional to x, calculate y when $x = 7$ if, when $x = 2$, y is
a. 1 *b*. 2 *c*. 7 *d*. -5 *e*. 0.5

9.2 Which of the following functions are of the type $y = ax$? $y = ax^2$? Where possible, fill in the missing numbers.

a.

y	0	5	10	15	—
x	0	1	2	3	4

b.

y	2	8	18	—
x	1	2	3	4

c.

y	0	3	6	9	11
x	0	1	2	3	—

d.

y	1	3	5	7	—
x	0	1	2	3	4

9.3 In which of the following is y inversely proportional to x? directly proportional to x? neither?

a.

y	2.0	1.0
x	1.0	2.0

b.

y	1.0	2.0
x	1.0	2.0

c.

y	4	8	-2
x	-2	-1	4

d.

y	3.0	5.0
x	0.0	2.0

9.4 Obtain an expression for $y_2 - y_1$ if

a. $y = ax$

b. $y = a/x$

c. $y = ax + b$

d. $y = ax^{1/2}$

e. $y = a/x^{1/2}$

9.5 Consider the equation

$$y = ax^{-1/2}$$

a. Calculate the ratio y_2/y_1 if $x_2 = 2.0$, $x_1 = 1.0$.

b. If $y_2 = 2y_1$ and $x_1 = 1.0$, what is the value of x_2?

9.11 The vapor pressure, P, of the solvent in a liquid solution is directly proportional to its mole fraction, X. Complete the following table:

P	____	____	____	____	160 mm Hg	____
X	0.00	0.20	0.40	0.60	0.80	1.00

9.12 Consider the reaction $A \rightarrow$ products, for which the rate expression is

$$\text{rate} = k(\text{conc. } A)^n$$

Determine the value of n if, when the concentration of A is doubled, the rate

a. doubles

b. increases by a factor of 4

c. remains unchanged

9.13 For the equilibrium between atomic fluorine, F, and molecular fluorine, F_2, the following data apply:

[F]	0.316	0.447	0.548
$[F_2]$	1.00	2.00	3.00

The functional relationship is of the form $[F_2] = K\,[F]^n$, where n is a positive, whole number (1, 2, 3, . . .). Use the data to obtain n and then K.

9.14 The average velocity of a gas molecule, u, is directly proportional to the square root of the absolute temperature, T, and inversely proportional to the square root of the molecular mass, M.

a. Express these two relationships in terms of a single equation (use a to represent the constant).

b. Write a two-point equation relating u_2 to u_1, T_2, T_1, M_2, and M_1.

9.15 The rate of effusion of a gas is inversely proportional to the square root of its molecular mass.

a. Calculate the ratio of the rates of effusion of CH_4, ($M = 16.0$) against He($M = 4.00$).

b. Calculate the molecular mass of a gas which effuses 0.82 times as rapidly as O_2 ($M = 32.0$).

9.6 Given the equation

$$y = -61.5 - kx$$

suppose that y is 16.0 when x is 8.0. What is the value of

a. k *b*. y when $x = 4.0$

c. x when $y = 0$

9.7 Obtain the two-point equation corresponding to

a. $\log y = -ax + b$

b. $y = -a \log x$

9.8 Using the equation

$$\log y = \frac{-2.00}{x} + 2.00$$

complete the following table:

y	____	____	3.00
x	1.00	2.00	____

9.9 Evaluate the constants a, b, and c in the following equations:

a. $y = a/x$; $y = 6$ when $x = 4$

b. $y = ax + b$; $y = 6$ when $x = 4$, $y = 2$ when $x = 0$.

c. $y = a + bx + cx^2$; $y = 1$ when $x = 0$, $y = 2$ when $x = 1$, $y = 1$ when $x = 2$

9.16 For the reaction

$$2\ SO_2\ (g) + O_2\ (g) \rightarrow 2\ SO_3\ (g)$$

$\Delta H = -198.2$ kJ and $\Delta G = -140.0$ kJ at 300 K. Using the equation $\Delta G = \Delta H - T\,\Delta S$, calculate

a. ΔS *b*. ΔG at 500 K

c. the temperature at which $\Delta G = 0$

9.17 The equation relating the standard free energy change, $\Delta G°$, to the equilibrium constant, K, is

$$\Delta G° = -(2.303)(8.314)T \log K$$

Obtain a two-point equation for two reactions, expressing the difference in ΔG's ($\Delta G_2° - \Delta G_1°$), in terms of the equilibrium constants K_2 and K_1.

9.18 Consider the two-point equation relating the vapor pressure, P, of a liquid to the absolute temperature, T:

$$\log \frac{P_2}{P_1} = \frac{\Delta H_{vap}(T_2 - T_1)}{(2.303)(8.314)T_2T_1}$$

The vapor pressure for water is 760 mm Hg at 373 K. Taking ΔH_{vap} to be 4.70×10^4 J, calculate

a. the vapor pressure at 350 K.

b. the temperature at which the vapor pressure is 1000 mm Hg

9.19 Given the following data for the reaction

$$2\ NO(g) + O_2(g) \rightleftharpoons 2\ NO_2(g)$$

show that the ratio $[NO_2]^2/([NO]^2 \times [O_2])$ is constant, and evaluate the constant.

[NO]	$[O_2]$	$[NO_2]$
1.00	1.00	3.16
1.00	2.00	4.47
2.00	1.00	6.32
2.00	2.00	8.95

9.10 For each of the linear functions listed below, give the values of the constants a and b.

a.	y	3.6	6.0
	x	3.0	5.0
b.	y	0.5	3.0
	x	1.0	2.0
c.	y	6.0	3.5
	x	0.0	1.0

9.20 The volume, V, of one gram of mercury is a linear function of temperature, t:

$$V = at + b$$

Evaluate the constants in this equation, using the following data,

V (cm^3)	0.073688	0.074089
t(°C)	10.00	40.00

CHAPTER 10
Graphs

We saw in Chapter 9 that a functional relationship between two variables, y and x, can take the form of an algebraic equation, a table of y and x values, or a graph of y plotted against x. In that chapter, our discussion focused on the first two forms. Here, we will consider graphical methods of describing functional relations. To begin with, we survey the various types of graphs that you will come across in general chemistry.

10.1 TYPES OF GRAPHS AND HOW TO READ THEM

All graphs found in a textbook are two-dimensional in the sense that they are printed on a piece of paper. However, in a more meaningful way, graphs can be classified as one-, two-, or three-dimensional, according to the number of variables involved.

One-Dimensional Graphs

The simplest type of graph is one in which only a single variable is plotted. Such graphs are one-dimensional, since only distances in one dimension, either horizontal or vertical, have meaning. Two examples of a common type of one-dimensional graph in general chemistry are shown in Figure 10.1. In both cases, a single variable (oxidation number in 10.1a, heat of formation in 10.1b) is plotted along a vertical line. Values assigned to various substances containing nitrogen are shown by arrows to the right of the line.

Graphs of this type are easy to interpret. We see, for example, from Figure 10.1a that the oxidation number of nitrogen in NH_3 is -3. From Figure 10.1b, it appears that the heat of formation of NH_3 is about -45 kJ/mol, since the arrow for that compound is located about one fourth of the way from -40 to -60 kJ/mol.

Sometimes, the second dimension in a one-dimensional graph carries at least a qualitative meaning. In Figure 10.2, we show a typical activation energy diagram for the reaction

$$A \rightarrow B$$

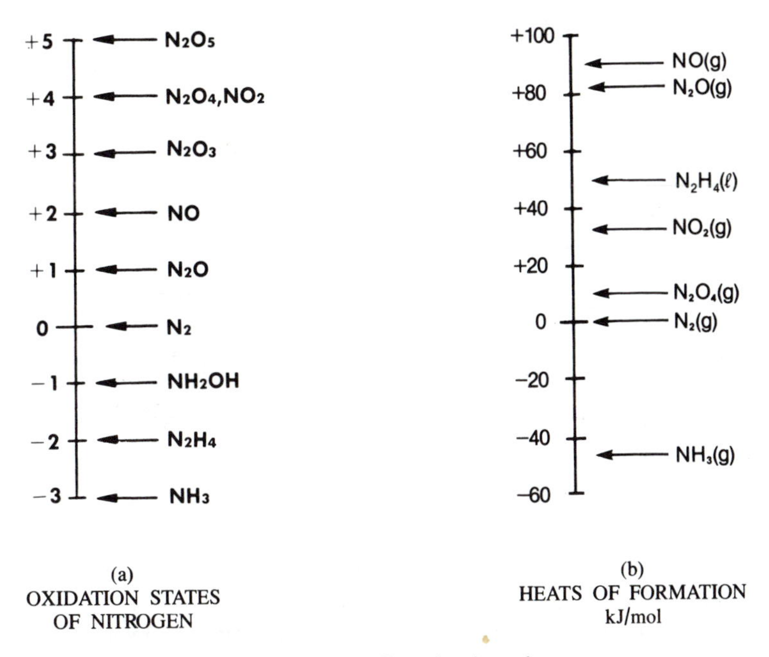

FIGURE 10.1 One-dimensional graphs.

The short horizontal line at the left shows the energy of the reactant, *A*. The line at the far right gives the energy of the product, *B*. The high point in the middle of the graph represents the energy of the "activated complex," *A**. This is an unstable intermediate formed in the course of the reaction. In a general way, distance in the horizontal direction represents time. We start with *A*, pass through the intermediate *A**, and arrive finally at *B*. However, horizontal distance has no quantitative meaning. We have no idea how long it takes to form *A** from *A* or *B* from *A*.

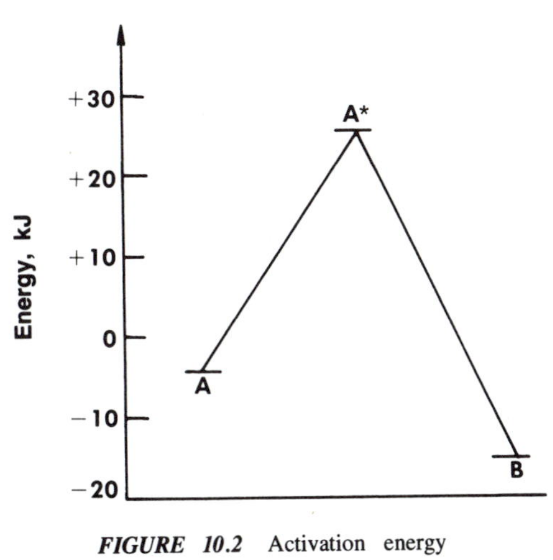

FIGURE 10.2 Activation energy diagram.

Example 10.1 The activation energy for the reaction shown in Figure 10.2 is defined as $E_{A*} - E_A$. Estimate the activation energy from the graph.

Solution

It appears that E_{A*} is about +25 kJ (approximately halfway between +20 and +30 kJ). Again, E_A appears to be about −4 kJ (a little less than halfway from 0 to −10 kJ). Thus,

$$\text{activation energy} = 25 \text{ kJ} - (-4 \text{ kJ}) = +29 \text{ kJ}$$

The uncertainty in reading the graph at both locations is of the order of ±1 kJ. Hence, we should not be surprised if our answer is off by as much as 2 kJ.

Two-Dimensional Graphs

By far the most common type of graph in chemistry (and the only type discussed in the remaining sections of this chapter) is two-dimensional. Such a graph shows the relationship between two variables, y and x. Points on the graph are located with respect to two different axes, perpendicular to each other (Figure 10.3). Note that

— along the horizontal axis the *abscissa* increases from left to right. Ordinarily, the abscissa is taken to be the independent variable, x. From Figure 10.3, we see that x increases from left to right at a uniform rate.

— along the vertical axis, the *ordinate* increases from bottom to top. Ordinarily, the ordinate is taken to be the dependent variable, y. As with x, the value of y changes at a uniform rate as we move along the axis.

— the point at which both x and y are zero is called the *origin* (marked as 0 in Figure 10.3). To the left of the origin, the value of x is negative; to the right, x is positive. Below the origin, y is negative; above the origin, y is positive.

A typical two-dimensional graph is shown in Figure 10.4. The x and y values corresponding to a point on such a graph are found in a straightforward way (Example 10.2).

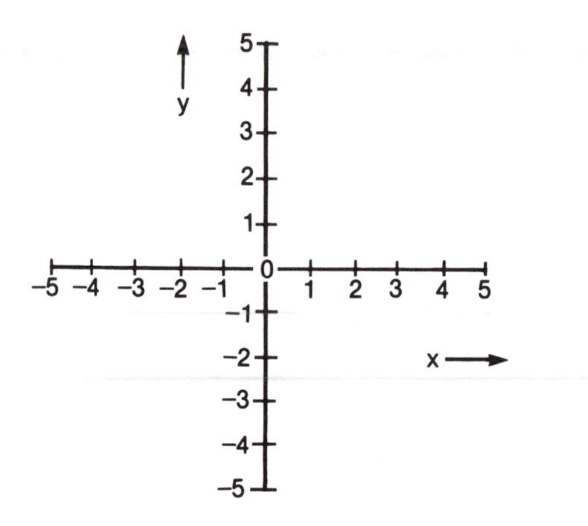

FIGURE 10.3 Vertical and horizontal axes for two-dimensional graphs.

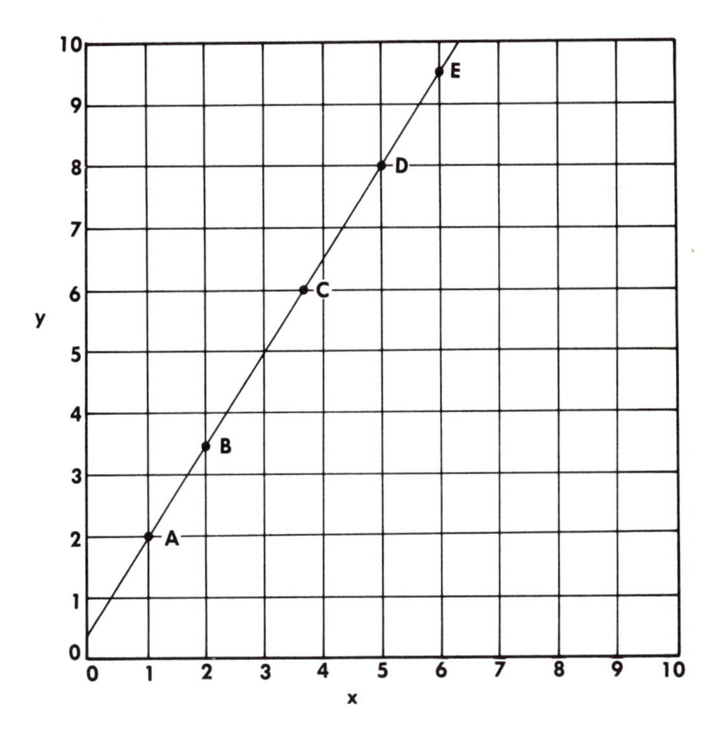

FIGURE 10.4 A simple straight-line graph.

Example 10.2 Referring to Figure 10.4, what are the x and y values of points

a. A b. B c. C

Solution

a. Point A lies at the intersection of
— the vertical line 1 unit to the right of the origin ($x = 1$)
— the horizontal line 2 units above the origin ($y = 2$).

$$\text{point } A: \quad x = 1, \quad y = 2$$

b. For point B, $x = 2$. The point appears to be about halfway from the line $y = 3$ to $y = 4$. A reasonable y value would be 3.5.

$$\text{point } B: \quad x = 2, \quad y = 3.5$$

c. For point C, y is 6. The point lies about two thirds of the way from the line $x = 3$ to $x = 4$. A reasonable estimate of the x value would be 3.7.

$$\text{point } C: \quad x = 3.7, \quad y = 6$$

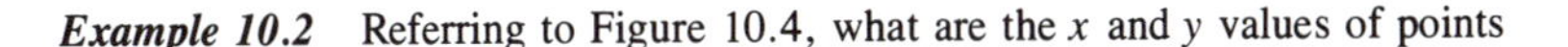

Most of the graphs found in a chemistry text differ from Figure 10.4 in one respect: the background grid of horizontal and vertical lines is omitted. Values of x and y are indicated simply by hash marks drawn in from the axes (Figure 10.5). This marking system has the advantage of making the graph stand out more clearly. It also makes it a bit more difficult to decide upon x and y values for a point.

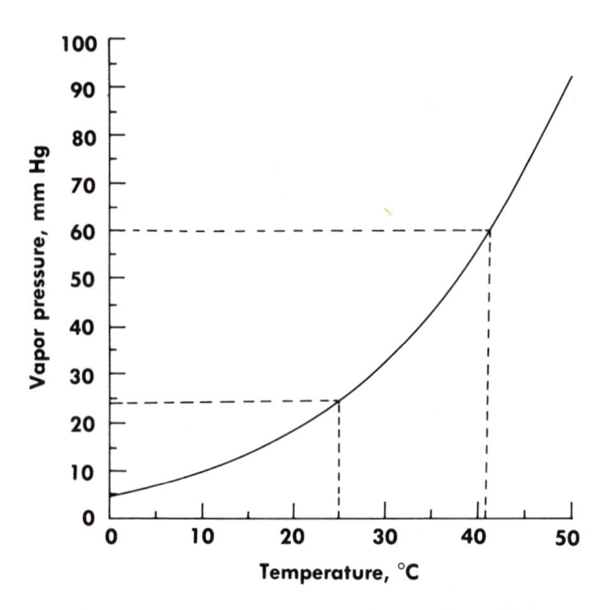

FIGURE 10.5 Vapor pressure of water, 0 to 50° C.

To assign x and y values to a point on a graph such as Figure 10.5, it helps to draw dotted lines from the point, perpendicular to the two axes. By observing where these lines intersect the axes, you can locate the point within fairly narrow limits (Example 10.3).

Example 10.3 Using Figure 10.5, which shows the vapor pressure of water as a function of temperature, estimate

a. the vapor pressure of water at 25°C.

b. the temperature at which the vapor pressure of water is 60 mm Hg.

Solution

Notice that along both axes the numbered marks are at intervals of 10 units (i.e., 10°C, 10 mm Hg). The small hash marks midway between the larger ones represent intervals of 5 units. For example, the mark midway between 20 and 30 on the horizontal axis represents 25°C.

a. We first draw a dotted vertical line up to the curve, starting at 25°C on the x axis. At the point where this line intersects the curve, we draw another line horizontally across to the y axis. The intersection of this horizontal line with the y axis appars to lie slightly below the hash mark at 25 mm Hg. A good estimate for the vapor pressure at 25°C would be 24 mm Hg.

b. Here, we reverse the procedure in (a). We start by extending the mark at 60 mm Hg across to the curve. Then we drop a perpendicular line down to the x axis. The intersection of this line with the axis lies slightly beyond 40°C, perhaps at 41°C.

Three-Dimensional Graphs

To show the functional relationship between three variables, x, y, and z, we could construct a solid, three-dimensional model. More commonly, graphs of this type are drawn in perspective on paper. The three axes x, y, and z are considered to be at 90° angles to each other. In Figure 10.6, we show the z axis running north and south. The x axis runs east and west; the y axis is perpendicular to the plane of the paper. A point drawn on such a graph represents one set of x, y, and z values. A group of such points determines a surface of a three-dimensional figure.

Graphs involving three variables are rather rare in general chemistry. One area where they are helpful is in describing the shapes of atomic orbitals. Figure 10.6 shows the three p orbitals (p_x, p_y, p_z). Each of the sets of twin lobes encloses a region of space in which there is a high probability of finding a p electron. This graph gives us little quantitative information. Indeed, we have not even attempted to show distances along the three axes. However, Figure 10.6 does give us valuable qualitative data. It tells us that the three p orbitals have identical shapes, are oriented at 90° angles to each other, and consist of two identical lobes, one on each side of the axis. This information can be helpful in making predictions about the electronic structures and geometries of molecules.

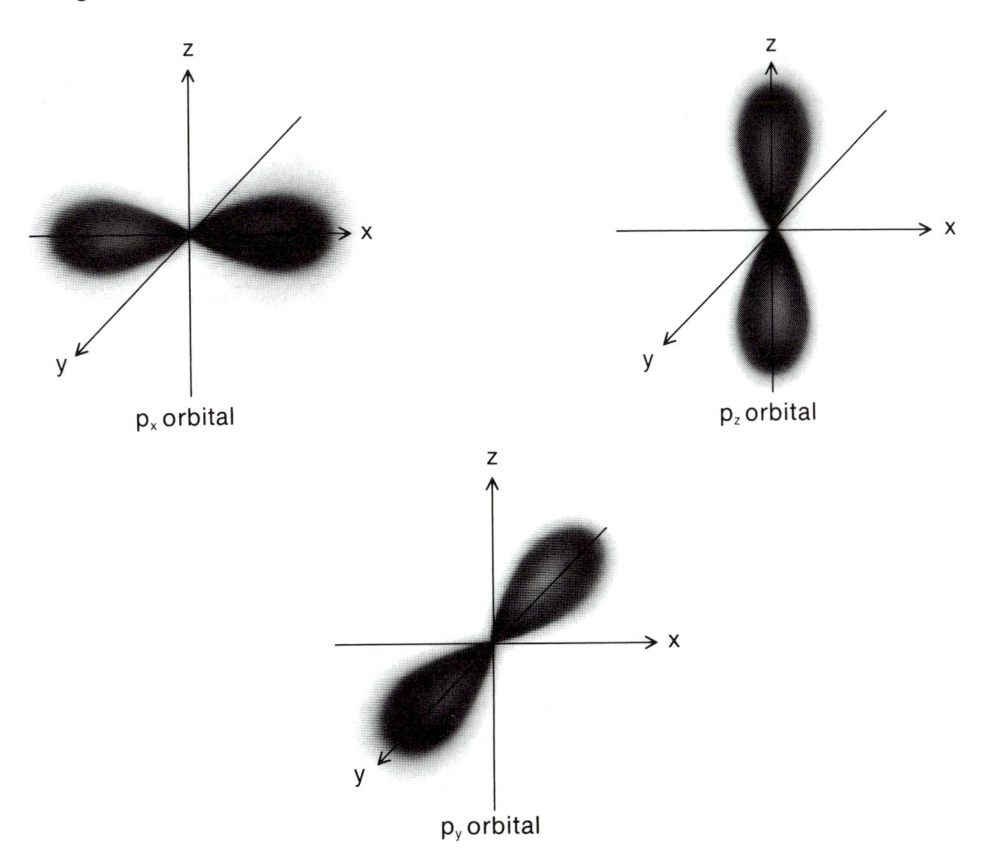

FIGURE 10.6 Electron clouds corresponding to the three p orbitals. The electron density in one of these orbitals is symmetrical about the x axis (p_x orbital). In another orbital, it is symmetrical about the z axis (p_z orbital) and in the third it is symmetrical about the y axis (p_y orbital). We describe this situation by saying that the three orbitals are directed at 90° angles to each other.

10.2 GRAPHING DATA

In Section 10.1, we learned how to read a two-dimensional graph. Basically, what we did was to generate a pair of x and y values corresponding to points on the graph. Here we consider the reverse problem: constructing a graph from data expressing y as a function of x. We will assume that we are working with a particular type of graph paper (Figure 10.7, p. 134) which has

— 10 major divisions (heavy lines) and 50 small divisions along the vertical axis.

— 7 major divisions and 35 small divisions along the horizontal axis.

There are two basic principles involved in drawing a graph. In order of priority, these are:

(1) The graph should be clear, easy to read, and easy to construct. It should be possible to determine x and y values of a point with a minimum of effort and see immediately what they represent (e.g., pressure in atmospheres, temperature in K).

(2) The graph should cover most of the graph paper rather than being squeezed into a tiny area. If the points on a graph are too close together, the graph loses much of its value.

To see how these principles work out in practice, we will construct a graph from the data in Table 10.1. This table gives the solubility of tartaric acid in water as a function of temperature. We will use the horizontal axis to show temperature, the independent variable. The dependent variable, solubility, will be shown on the vertical axis. In drawing this or any other graph, it is helpful to follow the logical sequence of steps outlined below.

Table 10.1 SOLUBILITY OF TARTARIC ACID (g/100 g WATER) AT VARIOUS TEMPERATURES (°C)

POINT	*SOLUBILITY*	*TEMPERATURE*
1	18	20
2	25	30
3	37	40
4	50	50
5	65	60
6	81	70
7	98	80

Steps in Drawing a Graph

1. DECIDE HOW MANY UNITS EACH DIVISION ALONG THE TWO AXES WILL REPRESENT. Let's consider the horizontal axis first. Remember that we have 7 large divisions along this axis. The temperatures in Table 10.1 run from 20°C to 80°C, a range of 60°. If we insisted upon making the graph cover the entire paper, the horizontal axis would start at 20° at the far left and go to 80° at the far right. This would require that each major division correspond to

$$\frac{60°}{7} = 8.57° \ . \ .$$

Each minor division (5 per major division) would represent 8.57/5 = 1.71 . . These numbers would be awkward to work with, to say the least. It makes a lot more sense to make each major division along the horizontal axis represent 10°. With that choice, each minor division becomes 10°/5 = 2°. Now it is relatively easy to read the temperature at a point on the graph.

On the vertical axis, we have to cover a range of 80 solubility units:

$$98 \text{ g/100 g water} - 18 \text{ g/100 g water} = 80 \text{ g/100 g water}$$

This must be done within 10 major divisions. We must assign to each major division at least

$$\frac{80}{10} = 8$$

solubility units. The graph will be somewhat easier to read if we make each major division along the vertical axis represent 10 units rather than 8. This way, each smaller division will represent 2 units rather than 1.6.

Let us summarize the reasoning process just described. We first divide the range to be covered (60°C, 80 solubility units) by the number of large divisions available along each axis (7, 10). The quotients obtained give us the *minimum number* of units that can be assigned to each major division. If we chose these values, we would make fullest use of the paper, spreading the graph over the total area. Ordinarily, by increasing these values somewhat, we arrive at a scale which makes the graph easier both to construct and to read.

2. DECIDE UPON THE MINIMUM VALUES OF x AND y, THOSE AT THE LOWER LEFT CORNER OF THE GRAPH. An obvious possibility here would be to put point 1 ($y = 18$, $x = 20$) at the lower left-hand corner. As a matter of fact, 20 is a good choice for the minimum x value. The major divisions along the x axis will then be 20°, 30°, 40°, 50°, 60°, 70°, 80°, and 90°. Each of the temperatures listed in Table 10.1 will then fall at a major division, a happy situation indeed.

On the other hand, 18 would not be a convenient starting point for y. The major divisions along the y axis would then read 18, 28, 38, . . . , 118. It would be easier to locate y values if we made the major divisions on the vertical axis correspond to integral multiples of 10. Perhaps the simplest way to do this is to choose 0 as the minimum y value. The major divisions on the y axis are now 0, 10, 20, . . . , 100. This covers neatly the range of solubilities, 18 through 98 g/100 g of water.

In general, to assign y and x values at the lower left-hand corner of the graph, we first look at the smallest y and x values in the table (e.g., 18, 20). Clearly, this point must be included on the graph. In the example we are discussing, we could not start the graph at $y = 20$, $x = 30$. Often, we will find it most convenient to start below the smallest set of y and x values (e.g., at 0, 20). However, we must be careful not to start so low that high values of y and x fall off the paper. Consider, for example, what would happen in this case if we were to assign $x = 0$, $y = 0$ to the lower left-hand corner. Keeping each major division at 10 units, the x axis would run from 0° to 70°. The highest temperature, 80°, would be pushed off the right edge of the paper. The point $y = 98$, $x = 80$ could not be plotted, an embarrassing situation.

3. NUMBER EACH MAJOR DIVISION ALONG BOTH AXES. If steps 1 and 2 have been carried out thoughtfully, this is routine. We write the numbers 0, 10, 20, . . . , at the left of the *y* axis, beside each major division. The numbers corresponding to major divisions on the *x* axis (20, 30, 40, . . .) are written below that axis.

4. LABEL BOTH AXES, INDICATING THE QUANTITY BEING PLOTTED. In doing this, be sure to indicate not only the identity of the variable (solubility, temperature), but also the units in which it is expressed (g/100 g water,°C).

In Figure 10.7 we show the results of Steps 1 to 4 for the case we are discussing. Note that the body of the graph paper is still untouched. Now, at long last, we are ready to construct the graph itself.

5. PLOT THE POINTS. Here we will consider only points 1 and 2. To find where point 1 belongs, we first locate "18" on the vertical axis. Clearly, it falls

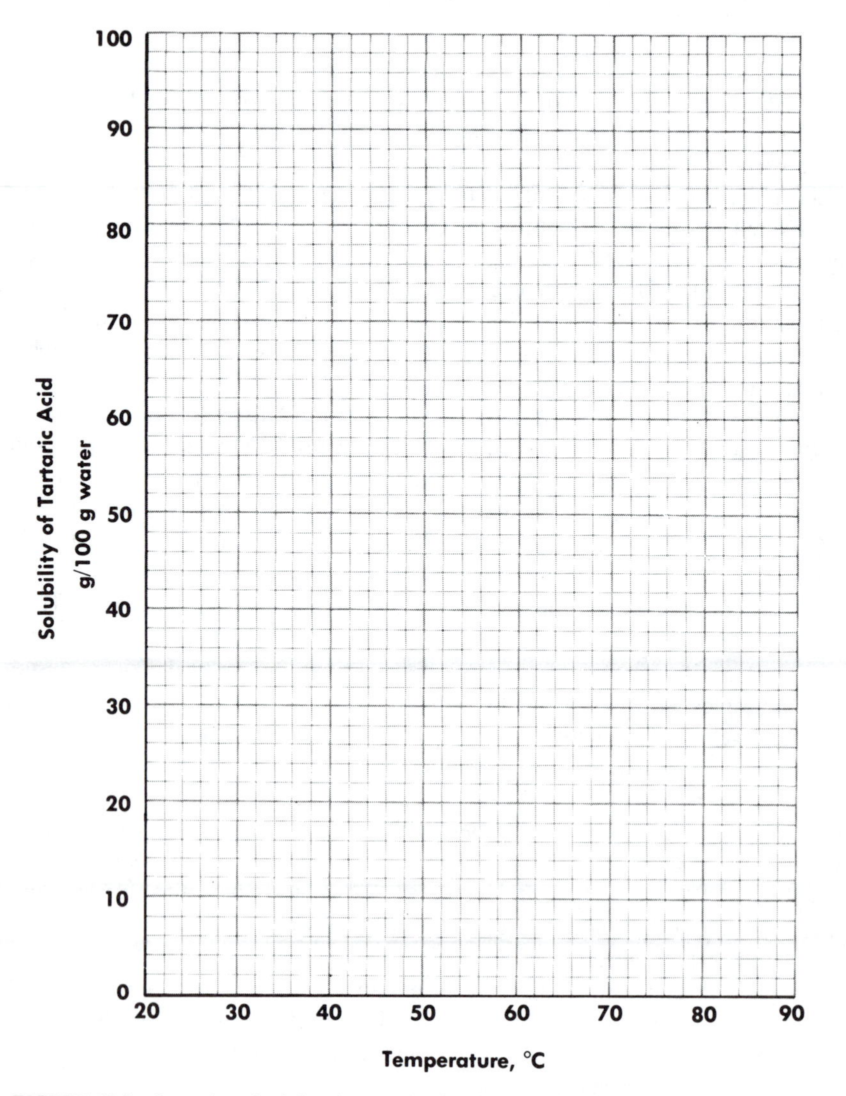

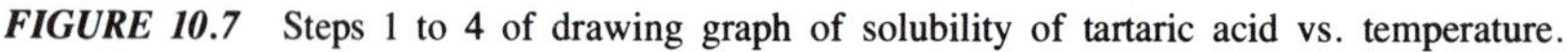

FIGURE 10.7 Steps 1 to 4 of drawing graph of solubility of tartaric acid vs. temperature.

between the major divisions labeled "10" and "20." To pin it down more closely, we note that each small division represents 2 solubility units. We conclude that a y value of 18 must fall one small division below 20. Keeping this in mind, we try to locate an x value of "20." As it happens, we don't have far to look. Since we started the x axis at 20°, any point for which $x = 20$ must fall on the heavy line at the left edge of the paper. In other words, the point $y = 18$, $x = 20$ falls on the y axis, 18 units up from the origin. This point is indicated as "1" on Figure 10.8.

To locate point 2, we first have to find where 25 falls on the y axis. Since 25 is midway between 20 and 30, it must be $2\frac{1}{2}$ small divisions above 20. You may find it helpful to enter a light hash mark at this point on the y axis. Now move out from this mark, parallel to the x axis, until you come to the vertical line corresponding to $x = 30$. This is one major division to the right of the vertical axis. At the intersection $y = 25$, $x = 30$, indicate point 2 with a dot.

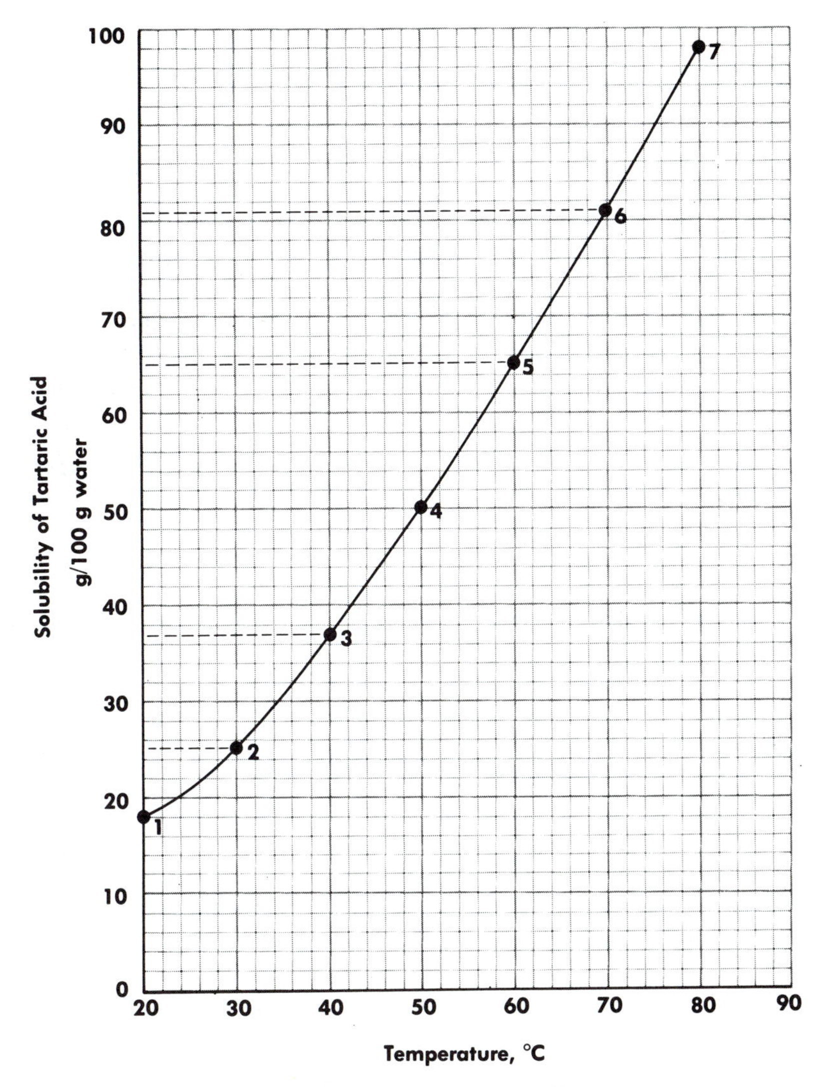

FIGURE 10.8 Graph of solubility of tartaric acid vs. temperature.

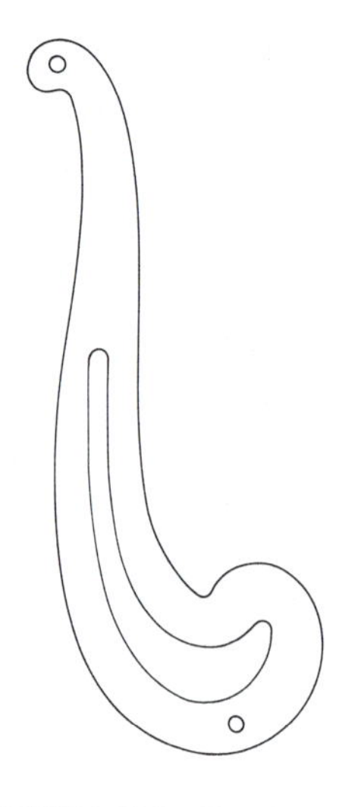

FIGURE 10.9 French curve.

Points 3 through 7 are located similarly. Point 4, for example, is located at the intersection of the lines $y = 50$, $x = 50$. You may wish to check that all of these points are located properly.

6. DRAW A SMOOTH CURVE THROUGH THE POINTS. For a straight line, this step is very easy indeed. Here, the plot is clearly not linear, at least in the lower portions. A French curve (Figure 10.9) or, even better, a flexible spline, is very helpful in drawing curves such as this. In using a French curve, avoid the common mistake of trying to draw too large a portion of the curve at once. You may have to shift the device several times to avoid getting sharp breaks in the curve.

Comments on Graphing

Sometimes, in drawing graphs to represent experimental data, you will run into problems not covered explicitly in the discussion above. A minor but annoying problem arises when the numbers involved are very large (e.g., 1,000,000) or very small (0.000 001). How should we label the major divisions along the axes in such cases? Writing out the numbers would consume a great deal of space, and the labels would run together and be hard to read. Two approaches may be used here:

(1) The numbers are expressed in exponential notation (e.g., 1×10^6, 1×10^{-6}). This is clear enough but doesn't save much space. Notice, for example, that there are still five characters in 1×10^6 and six in 1×10^{-6}.

(2) A number such as 1,000,000 is indicated on the axis simply as "1," thus solving the space problem. If this is done, the value of each division can be indicated by
— a statement written along the axis that "one division represents 10^6."
— labeling the axis as "$y \times 10^{-6}$." This is commonly done, and it can be very confusing, at least the first time you see it. The understanding is that

$$y \times 10^{-6} = 1, \quad \text{so } y = \frac{1}{10^{-6}} = 1{,}000{,}000$$

In this system, an x value of 0.000 001 would be indicated on the x axis as "1." The x axis itself would then be labeled as "$x \times 10^6$."

$$x \times 10^6 = 1, \quad \text{so } x = \frac{1}{10^6} = 0.000\ 001$$

In graphing data, it is desirable to indicate in some way the precision of the measurements involved. One way to do this is to adjust the size of the dots representing points on the graph. Small dots suggest data of high precision; large dots indicate rather crude data. A simpler approach is to indicate the number of significant figures in the measurements by the way in which divisions along an axis are numbered. Consider, for example, Figure 10.8. If the solubilities were known to ± 0.1 g, we could write the number 10.0, 20.0, . . . along the y axis instead of 10, 20, ...

With data of low precision, one or more points often appear to be far off the curve. This effect is most apparent with a straight-line graph (Section 10.4). It is less obvious but equally serious if the plot is curved. If there is good reason to believe that a particular measurement is in error, you can ignore that point in constructing the graph. Otherwise, use all the points and draw a smooth curve that comes as close as possible to each point. In general, you should *not* make a "zigzag" plot in an attempt to force the curve to pass through every point.

10.3 GRAPHING EQUATIONS

To graph an algebraic equation, we proceed as described in Section 10.2, adding one more step. We first use the equation to generate a set of data points (Example 10.4).

Example 10.4 Given the equation

$$°F = 1.8(°C) + 32°$$

construct a plot of °F (y axis) vs °C (x axis). Cover the range from 0 to 100°C.

Solution

Along the x axis, we start with 0°C at the left and let each major division count 20°. With this system, it seems reasonable to calculate Fahrenheit temperatures corresponding to 0°, 20°, 40°, 60°, 80°, and 100°C. Using the equation given, we arrive at the following set of data points:

°F	32	68	104	140	176	212
°C	0	20	40	60	80	100

We must now decide how to assign numbers to the major divisions along the y axis. Note that the range is 212° − 32° = 180°. This means that each of the major divisions must represent at least 180°/10 = 18°. To make graphing simpler, we increase this number to 20°. Our first impulse might be to let the minimum value of y, like that of x, be 0°. However, this choice clearly will not work. The top line on the paper would then represent 200°F; the highest temperature, 212°F, would fall off the grid. A simple way to avoid this problem is to start at 20° on the y axis. The heavy lines corresponding to major divisions along that axis will then run from 20° to 220° at 20° intervals.

At this point, you need only plot the points in the usual way and draw a smooth curve through them. The results are shown in Figure 10.10. Note that the "smooth curve" is actually a straight line.

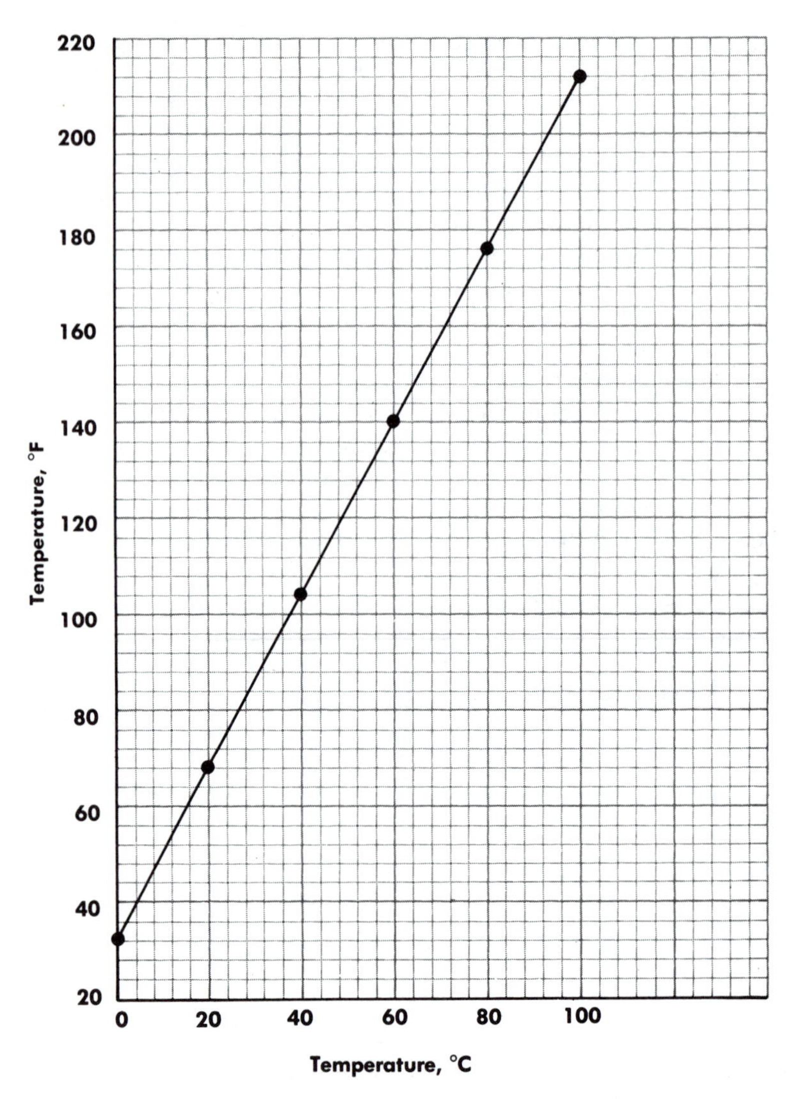

FIGURE 10.10 Graph of °F vs. °C from 0 to 100°C.

As Example 10.4 implies, in graphing an equation we usually need to generate only one data point per major division along the x axis. Sometimes, we can get by with still fewer points. If we are graphing a linear function of the form

$$y = ax + b$$

the situation is particularly simple. The position of a straight line is determined by two points. Hence, we only need to generate points at the beginning and end of the line. In Example 10.4, the two points

°F	32	212
°C	0	100

would have been sufficient to locate the line.

With certain equations we may find it helpful to generate more data points than usual. These points may be needed to locate the curve in a particular region (Example 10.5 and Figure 10.11).

Example 10.5 For one mole of an ideal gas at 92°C, the relation between volume and pressure is

$$V = \frac{30.0}{P} \qquad (V \text{ in liters, } P \text{ in atmospheres})$$

Plot V (y axis) against P (x axis) from $P = 3$ atm to $P = 10$ atm.

Solution

With 7 major divisions along the x axis, it is convenient to let each division represent 1 atm. We then calculate V at pressure intervals of 1 atm from 3 to 10 atm.

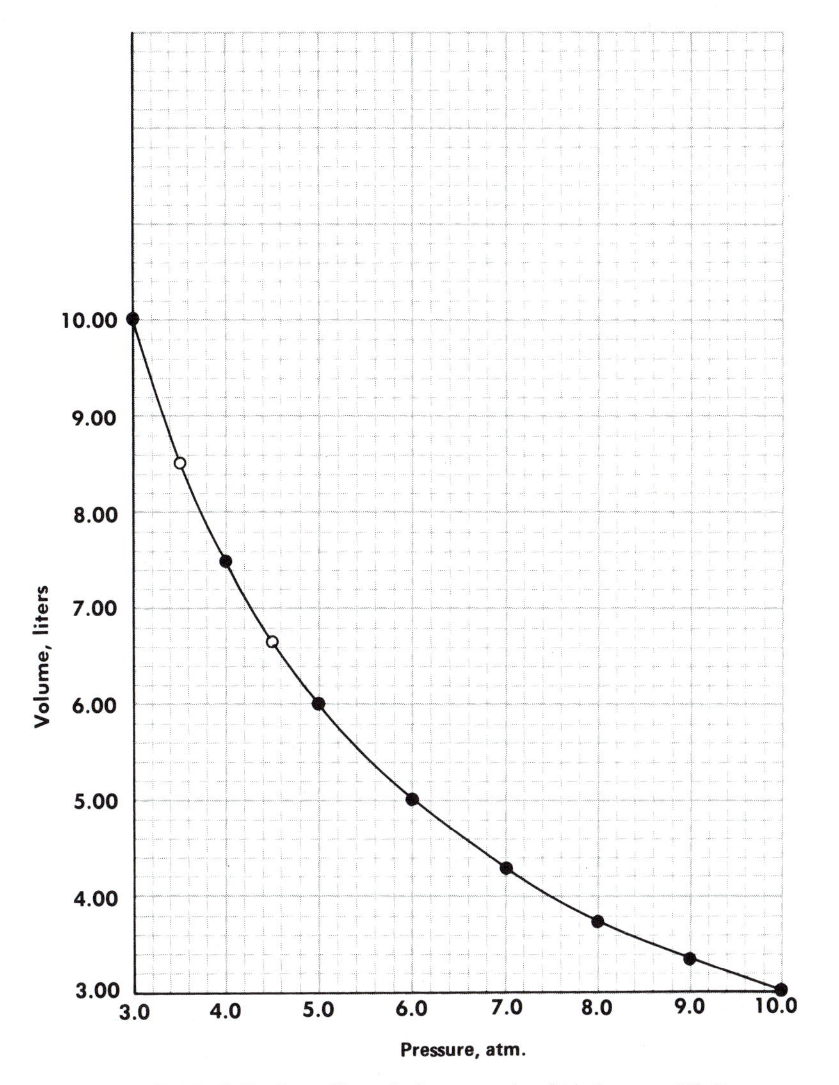

FIGURE 10.11 P vs. V graph for one mole of ideal gas at 92° C.

V (liters)	10.0	7.50	6.00	5.00	4.29	3.75	3.33	3.00
P (atmospheres)	3.0	4.0	5.0	6.0	7.0	8.0	9.0	10.0

We note that the values of V cover a range of $10.0 - 3.0 = 7.0\ \ell$. It seems reasonable here to make the y axis symmetric with the x axis. That is, we start with $V = 3\ \ell$ at the bottom and let each major division represent $1\ \ell$.

In Figure 10.11, the points generated above are plotted as solid dots. As you can see, V is decreasing rapidly near the beginning of the curve. This is particularly true in the region between 3 and 4 atm. To construct an accurate curve, it would help to have another point in this region, perhaps at 3.5 atm. Just to be safe, we might obtain another point at 4.5 atm.

$$P = 3.5 \text{ atm}; \quad V = \frac{30.0}{3.5} = 8.57\ \ell$$

$$P = 4.5 \text{ atm}; \quad V = \frac{30.0}{4.5} = 6.67\ \ell$$

These two points are shown as open circles in Figure 10.11. They appear to fill in gaps in the plot, locating the curve more reliably in this region.

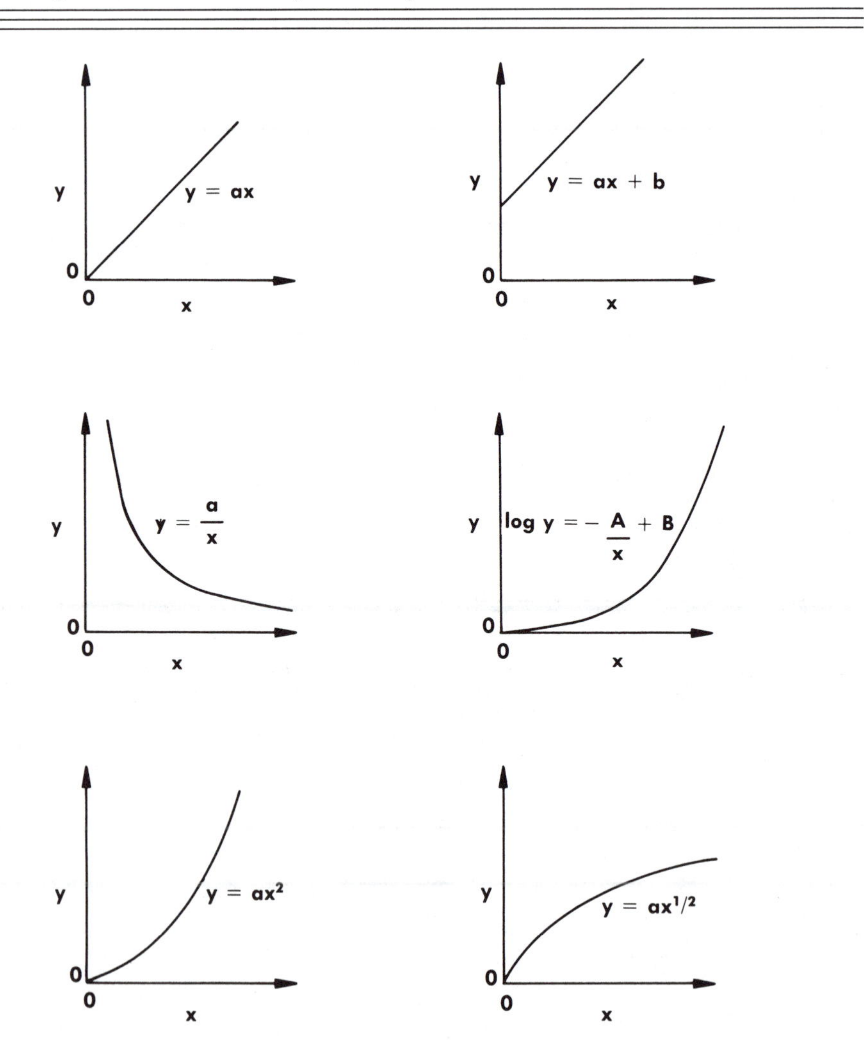

FIGURE 10.12 Typical graphs of functional relationships (all constants taken to be positive).

Often, in graphing an equation you are not expected to make an accurate plot on a piece of graph paper. A crude sketch indicating the general nature of the curve may be enough. In Figure 10.12, we show such sketches for various functional relationships discussed in Chapter 9. Two notes of caution are in order.

(1) In these sketches, only positive values of x and y are shown. Negative values are, of course, possible; recall Figure 10.3.
(2) In these graphs, the constants a and b were taken to be positive. The graphs could look quite different if a or b were negative (Section 10.4).

10.4 STRAIGHT-LINE GRAPHS

Most of the graphs that you will draw in general chemistry are of the straight-line type. As pointed out in Chapter 9, the general equation of a straight line is

$$y = ax + b \tag{10.1}$$

where a and b are constants. In this section we will look at some of the properties of straight-line graphs (linear functions).

Constants *a* and *b*

In Chapter 9, we showed that the two-point equation for a linear function is

$$\frac{y_2 - y_1}{x_2 - x_1} = a \tag{10.2}$$

But this quotient, $(y_2 - y_1)/(x_2 - x_1)$, is simply the ratio of the change in y (Δy) to the change in x (Δx). The quantity $\Delta y/\Delta x$ is, by definition, the slope* of the straight line. Hence, we see that the constant ***a* is the slope of the straight line obtained by plotting *y* vs. *x*.**

$$a = \text{slope} \tag{10.3}$$

To give a physical interpretation to b, we refer to Equation 10.1. Note that if $x = 0$, then $y = b$. This means that ***b* is the *y* intercept,** that is, the value of y where the straight line crosses the vertical axis corresponding to $x = 0$.

$$b = y \text{ intercept} \tag{10.4}$$

In Figure 10.13, p. 142 we show three straight lines. We read the y intercepts to be

Figure 10.13a: y intercept = 3, b = 3

Figure 10.13b: y intercept = 0, b = 0

Figure 10.13c: y intercept = 1, b = 1

*Note that by "slope" we do *not* mean the angle that the straight line makes with the x axis. Depending upon how we construct the graph, that angle could have a variety of different values, none of them meaningful. In contrast, $\Delta y/\Delta x = a$, can have only one value for a particular straight line (linear function).

The slopes ($\Delta y/\Delta x$ values) are not so obvious. To obtain them, it is probably best to work with the points at the ends of the lines. Thus we have

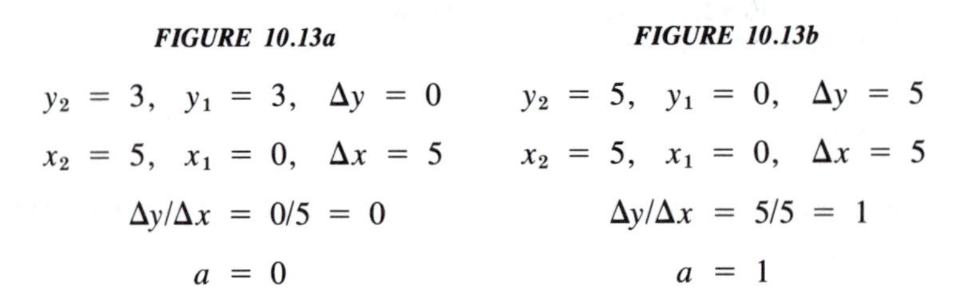

FIGURE 10.13a

$$y_2 = 3, \quad y_1 = 3, \quad \Delta y = 0$$
$$x_2 = 5, \quad x_1 = 0, \quad \Delta x = 5$$
$$\Delta y/\Delta x = 0/5 = 0$$
$$a = 0$$

FIGURE 10.13b

$$y_2 = 5, \quad y_1 = 0, \quad \Delta y = 5$$
$$x_2 = 5, \quad x_1 = 0, \quad \Delta x = 5$$
$$\Delta y/\Delta x = 5/5 = 1$$
$$a = 1$$

FIGURE 10.13c

$$y_2 = 3, \quad y_1 = 1, \quad \Delta y = 2$$
$$x_2 = 5, \quad x_1 = 0, \quad \Delta x = 5$$
$$\Delta y/\Delta x = -2/5 = 0.4$$
$$a = 0.4$$

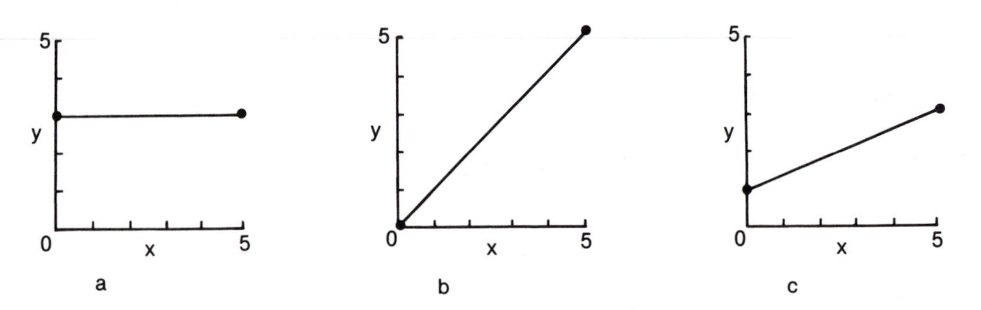

FIGURE 10.13 Three straight lines with different slopes and y intercepts.

Example 10.6 What are the equations of the three straight lines shown in Figure 10.13?

Solution

The general equation is

$$y = ax + b$$

Substituting the values of a and b obtained above,

Figure 10.13a: $y = 3$

In general, a straight line parallel to the x axis will have a slope of zero and so will correspond to the equation $y = b$

Figure 10.13b: $y = x$

In general, a straight line passing through the origin will have a y intercept of zero and so will correspond to the equation $y = ax$.

Figure 10.13c: $y = 0.4x + 1$

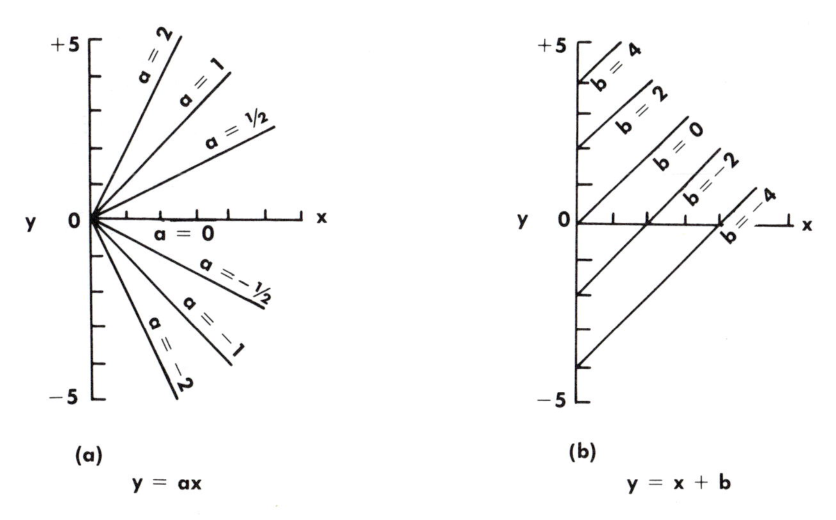

FIGURE 10.14 Effect of constants a (slope) and b (y intercept) on position of linear graph.

So far, all the straight lines we have dealt with have had positive (or zero) slopes and intercepts. However, negative values of a and/or b are quite common. Notice from Figure 10.14 that

— a negative value of a (negative slope) means that y decreases as x increases.

— a negative value of b (negative y intercept) means that the straight line intersects the y axis below the origin.

Conversion to Linear Functions

A variety of more complex functional relationships can be converted to linear functions by a simple change in variables. Consider, for example, the function

$$y = \frac{a}{x}$$

A plot of y vs. x is a hyperbola. A typical example of such a plot is the pressure-volume graph shown in Figure 10.11. However, if we plot y vs. $1/x$, we get a straight line with slope of a (Figure 10.15, p. 144). By changing the independent variable from x to $1/x$, we transform a hyperbola into a straight line.

Another functional relationship which is readily converted to a linear function is

$$\log y = \frac{-A}{x} + B$$

If we choose $\log y$ and $1/x$ to be our variables, rather than y and x, we obtain a straight line.

Converting a more complex functional relationship into a linear function has several advantages. For one thing, it greatly simplifies the process of **extrapolation**, that is, extending the graph to predict values of y beyond the range of data plotted. Suppose, for example, we wanted to use Figure 10.11 to find the volume at a pressure of 2.0 atm. This would be difficult, since it would require extending the curve in a

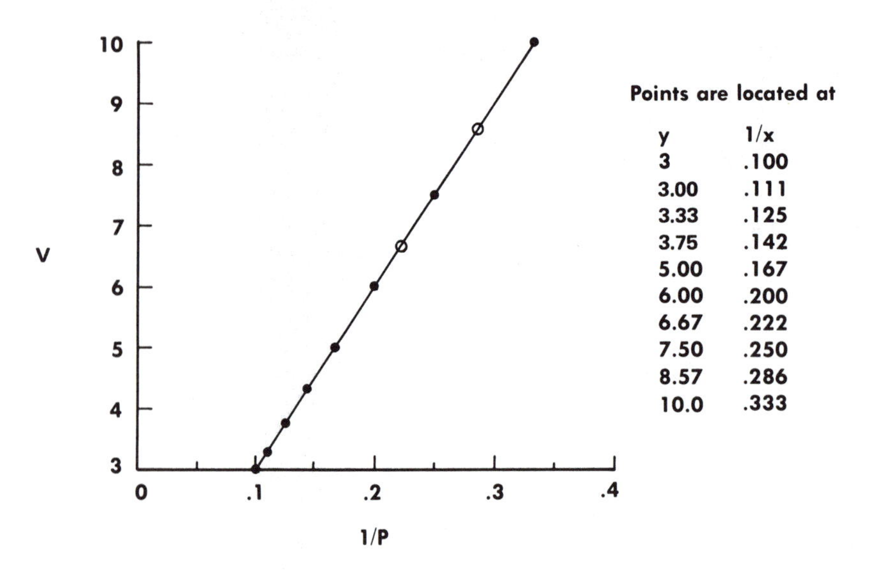

FIGURE 10.15 Conversion of hyperbola into a straight line. The points plotted are those shown in Figure 10.11.

region where the slope is changing rapidly. On the other hand, it is easy to extend the straight line in Figure 10.15. By doing so, we find

$$V = 15.0 \text{ when } P = 2.0 \quad (1/P = 0.50)$$

Interpolation (reading values of y between data points on the curve) is also easier with a straight line. There is always some doubt as to how a curve should be drawn through successive points, particularly if they are far apart. We have more confidence in our ability to locate the position of a straight line and hence in the values we read from it.

Converting complex functions to linear functions has still another advantage. It often happens that the slope (and sometimes the intercept) of the resulting straight line has direct physical meaning. Consider, for example, the relationship between the rate constant for a reaction, k, and the absolute temperature, T:

$$\log k = \frac{-E_a}{(2.303)(8.314)\,T} + B$$

The quantity E_a in this equation represents the activation energy in joules. Suppose now that we plot $\log k$ (y axis) vs. $1/T$ (x axis), as in Figure 10.16. Comparing this equation to the basic straight-line equation,

$$y = ax + b$$

we see that $\log k = y$, $1/T = x$, and that the slope must be the coefficient of the $1/T$ term. That is,

$$a = \text{slope} = \frac{-E_a}{(2.303)(8.314)} \tag{10.5}$$

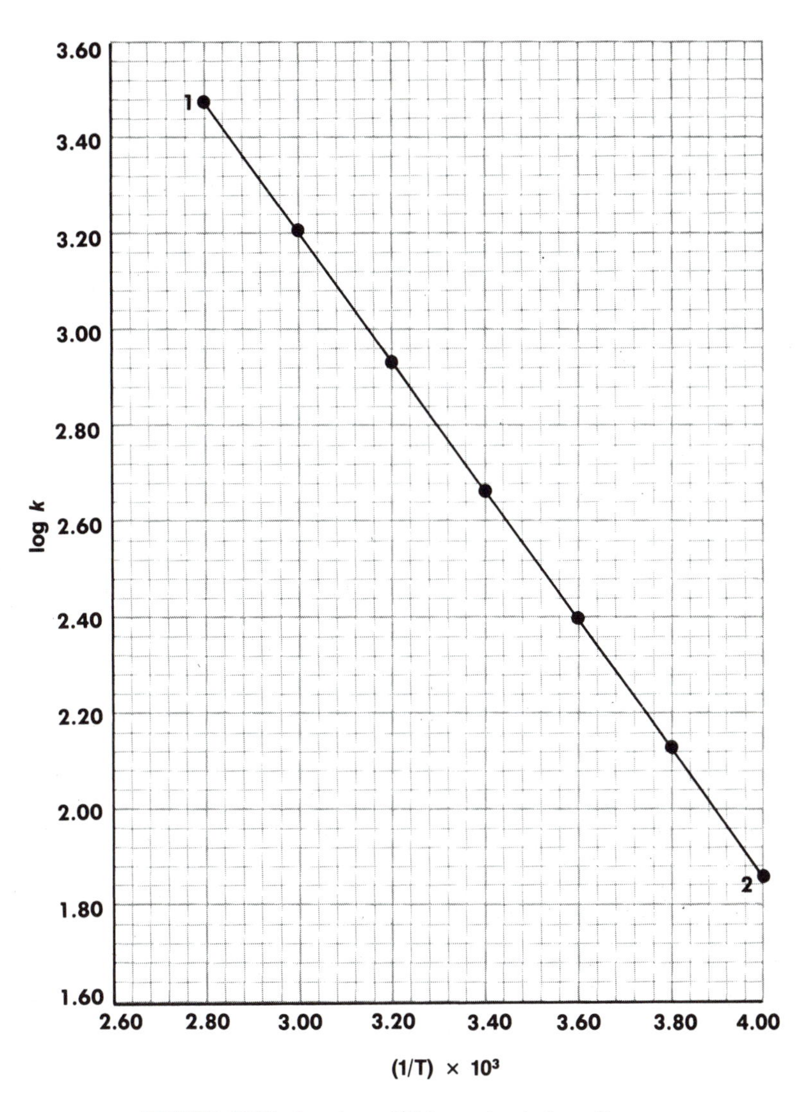

FIGURE 10.16 Log k vs. 1/T for a chemical reaction.

This equation allows us to calculate the energy of activation for a reaction from a plot of log k vs. $1/T$ (Example 10.7).

Example 10.7 Use the straight-line graph in Figure 10.16 to estimate the activation energy of the reaction.

Solution

The simplest way to evaluate the slope of this line is to work with the points at each end. For point 2,

$$x_2\text{:}\quad (1/T_2) \times 10^3 = 4.00;\quad 1/T_2 = 4.00 \times 10^{-3}$$

$$y_2\text{:}\quad \log k_2 = 1.86$$

For point 1:

$$x_1: \ (1/T_1) \times 10^3 = 2.80; \quad 1/T_1 = 2.80 \times 10^{-3}$$

$$y_1: \ \log k_1 = 3.48$$

Now we can evaluate the slope, $\Delta y/\Delta x$:

$$\text{slope} = \frac{y_2 - y_1}{x_2 - x_1} = \frac{1.86 - 3.48}{(4.00 - 2.80) \times 10^{-3}} = \frac{-1.62 \times 10^3}{1.20} = -1.35 \times 10^3$$

(Note that the slope is negative; y decreases when x increases.)

To evaluate the activation energy, we use Equation 10.5:

$$E_a = -(2.303)(8.314) \cdot \text{slope}$$

$$-(2.303)(8.314)(-1.35 \times 10^3) \text{ J} = 2.58 \times 10^4 \text{ J}$$

Drawing the "Best" Straight Line Through Data Points

Ordinarily, you will find it impossible to draw a straight line that passes *exactly* through all of the data points for a linear function. At best, some of the points will be above the line and some will be below it. In general, you should draw the line in such a way that there are about as many points above the line as below. It helps to use a transparent straightedge or triangle so you can see all the points as you draw the line.

Example 10.8 The following data are obtained for the free energy change of a reaction, ΔG, as a function of temperature:

ΔG (kJ)	2.8	4.8	6.0	7.7	9.4
T (K)	100	200	300	400	500

Use a graphical method to determine the constants ΔH (enthalpy change) and ΔS (entropy change) in the equation

$$\Delta G = \Delta H - T\,\Delta S$$

Solution

In Figure 10.17 we have plotted ΔG vs. T; the 5 data points are shown as small circles. We have attempted to draw the "best" straight line through these points. Notice that the points at 200 K and 500 K fall slightly above the line. Balancing this, the points at 100 K and 300 K fall slightly below the line.

Comparing the equation given above to the general linear function

$$y = ax + b$$

we see that the y intercept (b) is ΔH and the slope (a) is $-\Delta S$. We can read the y intercept from the graph as about +1.5 kJ. Thus,

$$\Delta H = +1.5 \text{ kJ}$$

To obtain the slope, we evaluate ΔG and T at the ends of the line.

$$\text{upper end: } \Delta G = +10.0 \text{ kJ}, \quad \text{T} = 550 \text{ K}$$

$$\text{lower end: } \Delta G = +1.5 \text{ kJ}, \quad \text{T} = 0 \text{ K}$$

$$\text{slope} = \frac{(10.0 - 1.5)\text{kJ}}{(550 - 0)\text{K}} = \frac{8.5 \text{ kJ}}{550 \text{ k}} = 0.015 \text{ kJ/K}$$

$$\Delta S = -\text{slope} = -0.015 \text{ kJ/K}$$

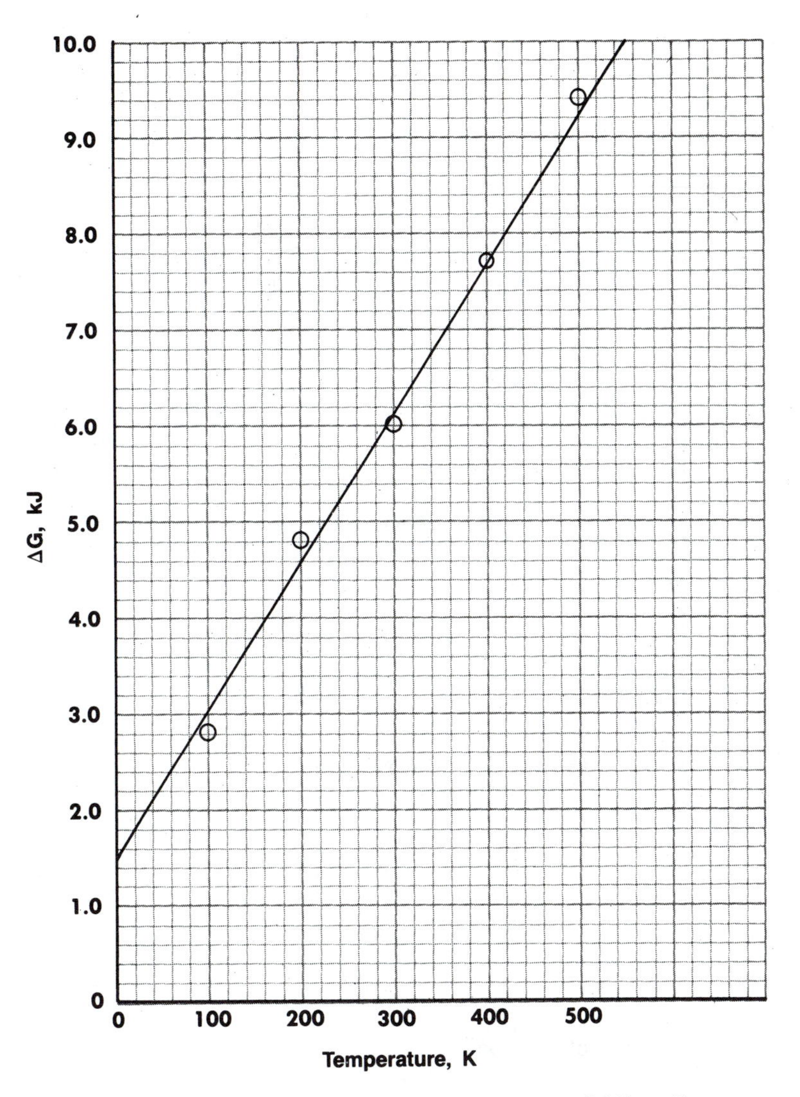

FIGURE 10.17 Best straight line through points of ΔG vs. T.

Drawing a straight line through a series of data points can be a somewhat uncertain process if considerable scatter is involved. There is an alternative way to determine the constants a and b in the equation

$$y = ax + b$$

This is referred to as the method of **least squares**. It leads to values of a and b such that the sum of the squares of the vertical distances of points from the line is as small as possible (Figure 10.18, p. 148). In this sense, least squares gives us the "best" values of a and b and hence the "best" straight line through the points.

We will not attempt to describe the theory behind this approach, which is based upon principles of calculus. Suffice it to say that the least squares method leads to two

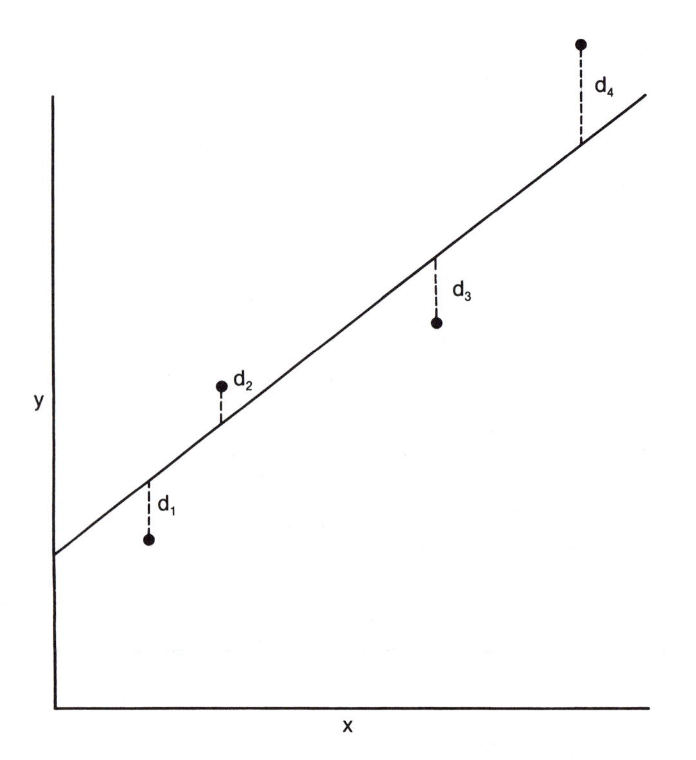

FIGURE 10.18 The "least squares" straight line is drawn so that $\Sigma d^2 = d_1^2 + d_2^2 + d_3^2 \div d_4^2 \div \cdots$ is a minimum.

equations which can be solved simultaneously for a and b. These equations are

$$\Sigma y = nb + a\Sigma x \tag{10.6}$$

$$\Sigma yx = b\Sigma x + a\Sigma x^2 \tag{10.7}$$

Here, n = number of data points

Σy = sum of all the y values = $y_1 + y_2 + y_3 + \cdots$

Σx = sum of all the x values = $x_1 + x_2 + x_3 + \cdots$

Σyx = sum of all the yx products = $y_1x_1 + y_2x_2 + y_3x_3 + \cdots$

Σx^2 = sum of all the x^2 values = $x_1^2 + x_2^2 + x_3^2 + \cdots$

Example 10.9 Consider the following data points:

y	1.0	3.1	4.9	6.8	9.1
x	0	1	2	3	4

Using the method of least squares, find the "best" values of a and b in the equation of the straight line

$$y = ax + b$$

corresponding to this data.

Solution

Since we have 5 data points, n in Equation 10.6 is 5. To show how Σy, Σx, Σyx and Σx^2 are obtained, it is helpful to set up a table.

y	x	yx	x^2
1.0	0	0.0	0
3.1	1	3.1	1
4.9	2	9.8	4
6.8	3	20.4	9
9.1	4	36.4	16
$\Sigma y = 24.9$	$\Sigma x = 10$	$\Sigma yx = 69.7$	$\Sigma x^2 = 30$

The two simultaneous equations must then be

$$24.9 = 5b + 10a$$

$$69.7 = 10b + 30a$$

To solve these equations, we follow the approach discussed in Chapter 8. We start by eliminating b. To do this, we multiply the first equation by 2 and subtract from the second:

$$69.7 = 10b + 30a$$

$$49.8 = 10b + 20a$$

$$19.9 = 10a; \quad a = 1.99$$

To find b, we substitute $a = 1.99$ in the first equation:

$$24.9 = 5b + 10(1.99)$$

$$5b = 5.0; \quad b = 1.0$$

The equation of the best straight line is then

$$y = 1.99x + 1.0$$

or, rounding off the constant a to 2 significant figures,

$$y = 2.0x + 1.0$$

Comparing this equation to the data points, we see that, of the y values,
— one ($y = 1$, $x = 0$) agrees exactly
— two are slightly high ($y = 3.1$, $x = 1$; $y = 9.1$, $x = 4$)
— two are slightly low ($y = 4.9$, $x = 2$; $y = 6.8$, $x = 3$)

With a calculator, the sums required for a least squares calculation can be found rapidly. Each of the quantities Σy, Σx, Σyx, and Σx^2 is obtained in one continuous operation. Indeed, some calculators have programs that allow you to simply enter successive values of y and x. Pressing the "least squares" keys then gives you directly the values of a and b.

PROBLEMS

10.1 Prepare a one-dimensional graph, similar to Figure 10.1, to show the 1970 populations of New York City (7.9×10^6), Chicago (3.4×10^6), Los Angeles (2.8×10^6), Philadelphia (1.9×10^6), Detroit (1.5×10^6), Houston (1.2×10^6), and Fall River (0.1×10^6).

10.2 Using Figure 10.4, give

a. the x and y values of points D and E.
b. y when $x = 3$; when $x = 4.5$
c. x when $y = 4$; when $y = 7$

10.3 Using Figure 10.10, estimate

a. °F when °C = 30; when °C = 72
b. °C when °F = 100; when °F = 200

10.4 Consider the data:

y	5.0	7.0	9.0	11.0	13.0
x	0.20	0.40	0.60	0.80	1.00

Plot this data and use the straight line obtained to estimate

a. y at $x = 0.30$ b. y at $x = 0.75$
c. x at $y = 10.0$

10.5 Draw graphs for the following equations in the regions indicated.

a. $y = 3x + 2$; $x = 0$ to $x = 6$
b. $y = 5/x$; $x = 1$ to $x = 6$

10.6 For the graph in Problem 10.4, determine the values of a and b in the equation $y = ax + b$.

10.11 For the reaction $A \rightarrow B$, E_a is 30 kJ; 25 kJ of energy is given off when A is converted to B. Taking A to have zero energy, construct an activation diagram similar to Figure 10.2.

10.12 Using Figure 10.8, estimate

a. the solubility at 25° C
b. the temperature at which the solubility is 60 g/100 g of water
c. the temperature at which the solubility is 100 g/100 g of water

10.13 Using Figure 10.17, estimate

a. the value of ΔG at 340 K
b. the value of ΔG at 0 K
c. the temperature at which $\Delta G = 5.0$ kJ

10.14 For the reaction

$$N_2(g) + 3\ H_2(g) \rightarrow 2NH_3(g)$$

the following data are obtained for the variation of the free energy change, ΔG, with temperature, T:

ΔG (kJ)	−32.8	−12.9	+7.0	+26.8	+46.7
T (K)	300	400	500	600	700

Plot the data and use the straight line obtained to estimate

a. ΔG at 350 K *b*. ΔG at 200 K
c. the temperature at which $\Delta G = 0$

10.15 For the reaction $2\ SO_2(g) + O_2(g) \rightarrow 2\ SO_3(g)$, ΔG (kJ) $= -198.2 + 0.1953\ T$. Plot ΔG vs. T from 300 to 1000 K.

10.16 Given that $\Delta G = \Delta H - T\Delta S$, where both ΔH and ΔS are taken to be constants, estimate ΔH and ΔS from the graph in Problem 10.14

10.7 Given the data

y	4.0	1.0	0.44	0.25
x	1.0	2.0	3.0	4.0

Plot y vs.

a. x *b*. $1/x$ *c*. $1/x^2$

Which of these is a straight line? What is the equation of the straight line?

10.8 Given the equation

$$y = 2x^2/z$$

plot

a. y vs. x at $z = 1$ from $x = 0$ to $x = 3$
b. y vs. z at $x = 1$ from $z = 1$ to $z = 3$

10.9 Given the data

y	1.00	3.16	4.65	5.62
x	1.00	2.00	3.00	4.00

a. complete the following table:

log y	____	____	____	____
$1/x$	____	____	____	____

b. plot log y vs. $1/x$.
c. from the plot in (*b*), determine the constant A in the equation

$$\log y = \frac{-A}{x} + 1.00$$

10.10 Given the data

y	−1.2	1.5	4.2	7.0	9.6	12.5
x	0.0	1.0	2.0	3.0	4.0	5.0

use the method of least squares to find the "best" values of a and b in the equation

$$y = ax + b$$

10.17 The following data apply for the concentrations of Ag^+ and CrO_4^{2-} in a solution saturated with Ag_2CrO_4:

conc. $CrO_4^{2-} \times 10^4$	1.0	0.25	0.11	0.067
conc. $Ag^+ \times 10^4$	1.0	2.0	3.0	4.0

a. Plot conc. CrO_4^{2-} (y) vs. conc. Ag^+ (x).
b. Plot conc. CrO_4^{2-} vs. $1/(\text{conc. } Ag^+)^2$.
c. Write the equation relating the concentration of CrO_4^{2-} to that of Ag^+.

10.18 Given the ideal gas law for one mole,

$$PV = RT$$

plot

a. P vs. T from 200 to 400 K at $V = 10\ \ell$ ($R = 0.0821\ \ell \cdot \text{atm/K}$)
b. PV vs. P from 1 to 7 atm at $T = 300$ K.

10.19 The following data were obtained for the vapor pressure, P, of chloroform as a function of temperature, T:

P	61.0	100.5	159.6	246.0
T	273	283	293	303

Proceed as in Problem 10.9 and obtain the heat of vaporization from the graph, using the equation

$$\log P = \frac{-\Delta H_{vap}}{(2.303)(8.314)T} + \text{constant}$$

10.20 Apply the method of least squares to the data in Example 10.8 to find the "best" values of ΔH and ΔS, and compare them to those obtained from the graph.

CHAPTER 11

Problem Analysis

Most of this book has been devoted to the mathematical techniques commonly used in general chemistry. We have stressed, particularly in the examples and end-of-chapter problems, how these techniques are applied in chemistry. We hope that this approach has helped you to sharpen your skills in mathematics. If it has also increased your understanding of chemical problems, so much the better.

In this final chapter we will not introduce any new mathematical techniques. Instead, we will concentrate upon the general subject of problem analysis. The objective of this chapter is to help you improve your ability to analyze and set up general chemistry problems. To do this, you must follow a logical path that leads from the written statement of a problem to its final solution. We start by describing a general approach that should prove useful in problem analysis.

11.1 GENERAL APPROACH

Hundreds of years ago, alchemists devoted their lives to a fruitless search for the "philosopher's stone" which would convert base metals to gold. By the same token, generations of students have searched for a "magic formula" which would allow them to solve any and all problems in general chemistry. Their efforts have been no more rewarding than those of the alchemists. There is no substitute for understanding what you are doing when you solve problems in chemistry.

Although there are no magic formulas for success, we can suggest a general approach to solving chemistry problems. This involves what amounts to a four-step analysis of the problem at hand. In this section, we will describe and illustrate each of these steps. Later, in Section 11.2, we will apply this approach to analyze a few typical problems in general chemistry.

1. READ THE PROBLEM CAREFULLY. MAKE SURE YOU KNOW WHAT IS GIVEN AND WHAT IS ASKED FOR.

Anyone who has graded examinations in general chemistry can cite cases where a student's answer did not correspond to the question asked. Presumably, the student failed to read the question carefully. A typical example, of a nonmathematical nature, is shown in Question A, Figure 11.1. Chemically, the answer given is correct. Had the student been asked to prepare NH_4Cl from NH_3, he would have deserved full credit. Unfortunately, he was asked the reverse question: how to prepare NH_3 from NH_4Cl.

Question A: How would you prepare ammonia, NH_3, from ammonium chloride, NH_4Cl?

Student Answer: Saturate a solution of hydrochloric acid with ammonia and evaporate to obtain crystals of NH_4Cl.

Question B: A sample of a certain gas occupies a volume of 210 cm³ at 300 K. At what temperature, in °C, will it have a volume of 250 cm³?

Student Answer: $T_2 = T_1 \times \frac{V_2}{V_1} = 300 \text{ K} \times \frac{250 \text{ cm}^3}{210 \text{ cm}^3} = 357 \text{ K}$

FIGURE 11.1

Another error, this one a bit more subtle, appears in Question B of Figure 11.1. Here again, everything the student has written is correct. She has related temperature and volume correctly, her arithmetic cannot be faulted, and indeed the final temperature *is* 357 K. However, the temperature was asked for in °C, not K. The student should have gone one step further and applied the relation

$$T \text{ (K)} = t \text{ (°C)} + 273$$

to obtain a final answer of 84°C.

Errors of the type just cited are easily avoided. It helps to *write down what is given in the statement of a problem and what you are asked to determine*. In Question B, for example, you might write

Given: initial volume (210 cm³), final volume (250 cm³), initial temperature (300 K)

Required: final temperature in °C

Writing out this information has other advantages. For one thing, it encourages you to analyze the problem carefully rather than starting to juggle numbers.

2. MAKE SURE YOU UNDERSTAND ALL THE TERMS USED IN THE STATEMENT OF THE PROBLEM.

Sometimes, you may be hung up on a problem because you don't understand a word or phrase used in its statement. The answer may hinge on the meaning of a particular term. This is the case with the multiple choice questions of Figure 11.2. These really involve little more than definitions. To answer them correctly, you must know what is meant by

— *atomic mass* (55.85 g Fe contains Avogadro's number—6.022×10^{23}—of atoms)

— *mole* (1 mol of Fe weighs 55.85 g)

Question 1: The atomic mass of iron is 55.85. The number of atoms in a sample of iron weighing 558.5 g is

a. 1 b. 10 c. 6.022×10^{24} d. 55.85 e. 558.5

Question 2: The number of moles of iron (Fe) in 558.5 g is

a. 0.1000 b. 1.000 c. 10.00 d. 55.85 e. 558.5

(The correct answer in each case is *c*)

FIGURE 11.2

Scientific terms used in a general chemistry text are usually defined in a glossary. This can be helpful if you come across a word that is unfamiliar to you. However, it is not enough to be able to rattle off definitions from memory. On homework assignments and examinations, you are seldom asked for definitions. Instead, you are expected to answer questions, such as those posed in Figure 11.2, that assume a working knowledge of chemical terms. There is a simple way to check the true extent of your chemical vocabulary. Try to explain orally what is meant by terms such as *atomic mass* or *mole*. When you can do this with no false starts or "hemming and hawing," you are ready for the most important step in problem analysis.

3. DECIDE WHAT CHEMICAL PRINCIPLE IS INVOLVED AND HOW YOU WILL USE IT TO SET UP THE PROBLEM.

Every problem in general chemistry involves applying one or more of the principles you have learned. These principles are stated and described in your text. In addition, your instructor will ordinarily discuss and illustrate them in class. You must become completely familiar with these principles before you attempt to work problems based upon them. Otherwise, you will not be able to recall the appropriate principle and apply it to the case in hand. This is the critical step in the analysis of any problem, chemical or otherwise.

Sometimes, as in Figure 11.3, a problem can be solved by applying a single principle in a straightforward way. In this case, you need only realize that *the coefficients in a balanced chemical equation represent the relative numbers of moles of each species.* Given the equation

$$2\ NO(g) + O_2(g) \rightarrow 2\ NO_2(g)$$

it should be clear that 2 moles of NO_2 are formed for every mole of O_2 that reacts (i.e., 1 mol O_2 $\simeq$ 2 mol NO_2, where the symbol $\simeq$ means "is chemically equivalent to"). Hence, if 1.43 mol of O_2 reacts, 2×1.43 or 2.86 mol of NO_2 must be produced.

Most problems in general chemistry do not unravel quite as simply as the one shown in Figure 11.3. Often, more than one principle is required to solve a particular problem. In that case, you must recognize each principle in turn and apply it in a logical way. As you may have discovered already, this is easier said than done. We will comment further on approaches to this type of problem in Section 11.2.

Many chemical principles are expressed in terms of algebraic equations (Chapter 7). Here, *you must understand clearly the meaning and the appropriate units for each symbol used in the equation.* As an example, consider one of the most common equations in

Problem: Given the balanced equation

$$2\ NO(g) + O_2(g) \rightarrow 2\ NO_2(g)$$

calculate the number of moles of NO_2 that can be produced from 1.43 mol of O_2.

Solution: 1 mol O_2 $\simeq$ 2 mol NO_2

$$\text{moles } NO_2 = 1.43 \text{ mol } O_2 \times \frac{2 \text{ mol } NO_2}{1 \text{ mol } O_2} = 2.86 \text{ mol } NO_2$$

FIGURE 11.3

general chemistry, the ideal gas law (Figure 11.4). If you take the gas constant R to be 0.0821 ℓ · atm/(mol · K), then, to be consistent, you must express

— pressure in atmospheres (*not* millimeters of mercury)
— volume in liters (*not* cubic centimeters)
— temperature in K (not in °C)

4. AFTER YOU HAVE ANALYZED A PROBLEM, SET IT UP IN A CONCISE, LOGICAL MANNER.

A great many problems in general chemistry can be handled by the "conversion factor" approach described in Chapter 2. This is illustrated in simple form in the set-up shown in Figure 11.3. Here the conversion factor 2 mol NO_2/1 mol O_2 is used to go from moles of O_2 to moles of NO_2.

At this stage, you may find it useful to review the material on unit conversions in Chapter 2. Remember that

— a conversion factor is a ratio of two quantities which are equal or equivalent to each other. Hence the conversion factor is itself equal to 1. Multiplying by such a factor does not change the value of a quantity, only its units.

— always carry units in working with conversion factors. This avoids a multitude of errors, most commonly that of getting the factor upside down.

— several conversion factors can be applied in succession. You may wish to do this in a several-step conversion. Alternatively, you may want to calculate in advance a single factor that will accomplish the entire conversion in a single step.

— in general: initial quantity × conversion factors(s) = desired quantity

If you use an algebraic equation to solve a chemistry problem, it is usually simplest to start by solving that equation for the quantity required. Suppose, for example, you are asked to calculate the number of moles of gas, n, in a sample, using the ideal gas law

$$PV = nRT$$

Here you might start by dividing both sides of the equation by RT to obtain

$$n = \frac{PV}{RT}$$

Now, substitute numbers for P, V, R, and T (in the proper units) and calculate the value of n.

The Ideal Gas Law

$$PV = nRT$$

P = pressure in atmospheres (1 atm = 760 mm Hg = 101.3 kPa = 14.7 lb/in^2)
V = volume in liters (1 ℓ = 1 dm^3 = 10^3 cm^3)
n = number of moles = (number of grams)/GMM, where GMM = mass of one mole in grams
R = 0.0821 ℓ · atm/(mol · K)
T = temperature in K (T (K) = t (°C) + 273)

FIGURE 11.4

Sometimes, in a several-step problem, you may analyze it correctly only to forget, halfway through, where you are going. To avoid such a disaster, i* is useful to write down the path you are following. Such a "route map" can guide you to your destination, the final answer. We shall consider a few such route maps in Section 11.2.

11.2 PROBLEM ANALYSIS; EXAMPLES

In this section we will apply the four-step system of problem analysis discussed in Section 11.1. To do this, we will consider a series of five examples, following the same basic approach in each case. We start with a relatively simple problem, but one which often gives students difficulty in the early stages of a general chemistry course.

Example 11.1 The atomic mass of sodium, Na, is 22.99. Calculate the mass in grams of a sodium atom.

Solution

(1) Given: atomic mass Na = 22.99 Required: mass (in grams) of one Na atom

(2) Two scientific terms are involved here, atom and atomic mass. An atom is the smallest particle of an element. The atomic mass is a number which expresses the average mass of an atom of an element relative to that of a C-12 atom (12.000 . .) Since sodium has an atomic mass of about 23, a sodium atom weighs about 23/12 as much as a C-12 atom.

(3) The principle involved here can be stated quite simply:

There are 6.022×10^{23} atoms in X grams of any element, where X is the atomic mass of the element. Applied to sodium, this leads to the relation

$$6.022 \times 10^{23} \text{ Na atoms} = 22.99 \text{ g}$$

This leads directly to the conversion factor we need to find the mass of a sodium atom.

(4) In effect, we want to convert 1 Na atom to grams. To do this, we use the conversion factor 22.99 g/6.022×10^{23} Na atoms. The set-up is

$$\text{mass} = 1 \text{ Na atom} \times \frac{22.99 \text{ g}}{6.022 \times 10^{23} \text{ Na atoms}} = 3.818 \times 10^{-23} \text{g}$$

The next problem resembles that worked in Figure 11.3. However, it is a bit more difficult. In particular, it requires that you understand just what is meant by a *mole*.

Example 11.2 Given the balanced equation

$$2 \text{ NO(g)} + O_2\text{(g)} \rightarrow 2 \text{ NO}_2\text{(g)}$$

calculate the mass in grams of NO_2 that can be produced from 2.46 mol of O_2.

Solution

(1) Given: 2.46 mol O_2
Required: number of grams of NO_2 formed when 2.46 mol of O_2 reacts

(2) The only term likely to give difficulty here is the mole (abbreviation mol). Look up the definition of a mole in your text. You will probably find that it represents 6.022×10^{23} particles, an interesting fact which is not very helpful here. More to the point, the mass of a mole in grams can be found by adding the atomic masses of the atoms in a formula. Taking the atomic masses of N and O to be 14.01 and 16.00 in that order, we have

$$1 \text{ mol } NO_2 = 14.01 \text{ g} + 2(16.00 \text{ g}) = 46.01 \text{ g}$$

(3) The principle here is that listed on p. 154. *The coefficients in a balanced equation represent the relative numbers of moles of each species.* Thus, we see that one mole of O_2 forms two moles of NO_2, or:

$$1 \text{ mol } O_2 \simeq 2 \text{ mol } NO_2$$

This relation allows us to determine the number of moles of NO_2 formed from 2.46 mol of O_2. Then, knowing that 1 mol of NO_2 weighs 46.01 g, we should be able to calculate the mass in grams of NO_2.

(4) Since two steps are involved, let's write down a "route map" rather than going directly to the set-up.
(a) Convert moles of O_2 to moles of NO_2 (1 mol $O_2 \simeq$ 2 mol NO_2)
(b) Convert moles of NO_2 to grams of NO_2 (1 mol NO_2 = 46.01 g NO_2)

Now, we set up the problem, using two conversion factors in succession:

$$\text{mass } NO_2 = 2.46 \text{ mol } O_2 \times \underset{\text{(a)}}{\frac{2 \text{ mol } NO_2}{1 \text{ mol } O_2}} \times \underset{\text{(b)}}{\frac{46.01 \text{ g } NO_2}{1 \text{ mol } NO_2}} = 226 \text{ g } NO_2$$

The two examples we have just worked involved the conversion factor approach. Now let's apply the four-step analysis to a problem which can be solved by using an algebraic equation, the ideal gas law.

Example 11.3 Determine the number of moles of hydrogen in a 10.0 ℓ cylinder at a pressure of 112 atm and a temperature of 20°C.

Solution

(1) Given: volume = 10.0 ℓ, pressure = 112 atm, temperature = 20°C
Required: number of moles of gas

(2) Here you need to know what is meant by the units of pressure, volume, and temperature. In particular, you must realize that temperature in °C can be converted to K by using the relation

$$T \text{ (K)} = t \text{ (°C)} + 273 = 20 + 273 = 293 \text{ K}$$

(3) The "principle" here is simply the ideal gas law

$$PV = nRT$$

where $R = 0.0821 \ \ell \cdot \text{atm/(mol} \cdot \text{K)}$

P = pressure in atmospheres

V = volume in liters

n = number of moles

T = temperature in K

Knowing the values of P, V, R, and T, we can calculate n.

(4) Solving the ideal gas law for n,

$$n = \frac{PV}{RT}$$

Substituting for P, V, R and T,

$$n = \frac{(112\ \text{atm})(10.0\ \ell)}{\left(0.0821\ \frac{\ell \cdot \text{atm}}{\text{mol} \cdot \text{K}}\right)(293\ \text{K})} = 46.6\ \text{mol}$$

Probably the most difficult type of problem in general chemistry is one in which two or more principles must be applied. Example 11.4 illustrates a rather simple case of this type.

Example 11.4 A sample of hydrogen gas is collected over water at 25°C (vapor pressure of water = 24 mm Hg). The total pressure of the wet gas is 752 mm Hg; its volume is 19.6 ℓ. Calculate the number of moles of hydrogen in the sample.

Solution

(1) Given: P_{tot} = 752 mm Hg; vapor pressure H_2O = 24 mm Hg; V = 19.6 ℓ; T = 25 + 273 = 298 K

Required: number of moles of hydrogen

(2) The only new term introduced here is vapor pressure. To work this problem, you must realize that the vapor pressure, 24 mm Hg, represents the pressure exerted by the water vapor mixed with the hydrogen gas.

(3) There are two principles involved, both of which are expressed by algebraic equations. One of these is the ideal gas law, referred to previously. Applied here, it takes the form

$$n_{H_2} = \frac{P_{H_2} \times V}{RT}$$

where P_{H_2} is the partial pressure of the dry hydrogen, that is, the pressure that the hydrogen would exert if it occupied the entire volume (19.6 ℓ) by itself at the same temperature (298 K).

To calculate P_{H_2}, it is necessary to use Dalton's law. This law states that the total pressure of a gas mixture is the sum of the partial pressures of the components of the mixture. Applied here, the resulting equation is

$$P_{tot} = P_{H_2} + P_{H_2O}$$

(4) We might start by applying Dalton's law to calculate P_{H_2}:

$$P_{H_2} = P_{tot} - P_{H_2O} = 752\ \text{mm Hg} - 24\ \text{mm Hg} = 728\ \text{mm Hg}$$

Recall (Figure 11.4), that if we are to use R = 0.0821 $\ell \cdot$ atm/(mol $\cdot$ K), pressure must be expressed in atmospheres. To change from millimeters of mercury to atmospheres, we use the conversion factor 1 atm/760 mm Hg:

$$P_{H_2} = 728\ \text{mm Hg} \times \frac{1\ \text{atm}}{760\ \text{mm Hg}} = 0.958\ \text{atm}$$

Now we are ready to solve for the number of moles of hydrogen:

$$n_{H_2} = \frac{P_{H_2} \times V}{RT} = \frac{(0.958\ \text{atm})(19.6\ \ell)}{\left(0.0821\ \frac{\ell \cdot \text{atm}}{\text{mol} \cdot \text{K}}\right)(298\text{K})} = 0.767\ \text{mol}$$

In practice, students seldom have much difficulty with the type of problem analyzed in Example 11.4. Perhaps this is because the two principles involved (ideal gas law, Dalton's law) are considered in the same chapter of the text and often discussed in the same lecture. When the different principles required to analyze a problem are more separated in space and time, the problem becomes considerably more difficult (Example 11.5).

Example 11.5 A certain hydrocarbon gas has the simplest formula C_2H_5. A sample of the gas occupying 0.224 ℓ at 1.00 atm and 27°C weighs 0.527 g. What is the molecular formula of the hydrocarbon?

Solution

(1) Given:

$$\text{simplest formula} = C_2H_5; \quad V = 0.224\ \ell; \quad P = 1.00\ \text{atm};$$
$$\text{g} = \text{mass of gas} = 0.527\ \text{g}; \quad T = 273 + 27 = 300\ \text{K}$$

Required: molecular formula

(2) The new terms introduced here are "simplest formula" and "molecular formula." The simplest formula gives the simplest atom ratio (2 C atoms per 5 H atoms). The molecular formula gives the composition of the molecule. It could be C_2H_5 or some integral multiple of C_2H_5, such as C_4H_{10}, C_6H_{15}, ...

(3) *Three* different principles are involved here:

(a) The number of moles of gas can be calculated from the ideal gas law.

$$n = \frac{PV}{RT}$$

(b) The molar mass in grams, GMM, can be obtained from the relation:

$$n = \text{g/GMM}$$

where g is the mass of the sample in grams.

(c) The molecular formula can be determined from the simplest formula, knowing the molar mass in grams. The "formula mass" in grams of C_2H_5 is 2(12.0 g) + 5(1.0 g) = 29.0 g. If the calculated molar mass turns out to be 29.0 g, the molecular formula must be the same as the simplest formula, i.e., C_2H_5. If it is 58.0 g, the molecular formula is twice the simplest formula, i.e., C_4H_{10}, and so on.

(4) With these principles in mind, a reasonable route map would be

(a) Calculate the number of moles of gas ($n = PV/RT$).

(b) Calculate the molar mass in grams (GMM = g/n).

(c) Determine the molecular formula, following the reasoning described above.

(i) $$n = \frac{PV}{RT} = \frac{(1.00\ \text{atm})(0.224\ \ell)}{\left(0.0821\ \frac{\ell \cdot \text{atm}}{\text{mol} \cdot \text{K}}\right)(300\ \text{K})} = 9.09 \times 10^{-3}\ \text{mol}$$

(ii) GMM = g/n = 0.527 g/(9.09 × 10^{-3} mol) = 58.0 g/mol

(iii) Since the molar mass is twice the formula mass (i.e., 58.0/29.0 = 2), the molecular formula must be C_4H_{10}.

It is unlikely that you could get very far with the analysis of Example 11.5 unless you had all three principles listed under (3) clearly in mind. One difficulty is that they are not ordinarily discussed together in the general chemistry course. The second and third principles, (b) and (c), are usually introduced early, perhaps in the first couple of weeks of the course. The ideal gas law, (a), usually comes later. By that time, many students have forgotten how to obtain the molecular formula from the simplest formula (c). They may even be unaware of relation (b), although this is really a defining equation for the mole.

The chances are good that, whenever you come across a problem that baffles you, more than one principle is involved. An effective way to analyze such a problem is to carry out what might be called a "reverse analysis." Ask yourself what piece of information you need to get the answer directly. Then consider how you might get that information from data in the problem. Several steps may be required as you work your way backward from the required quantity to those that are stated directly in the problem. In Table 11.1, we show such an analysis applied to Example 11.5.

Table 11.1 REVERSE ANALYSIS OF A MULTI-PRINCIPLE PROBLEM: (EXAMPLE 11.5)

REASONING:
(c) I could get the molecular formula from the simplest formula if I knew the molar mass.
(b) Since I know the mass of the sample, 0.527 g, I could find the molar mass if I knew the number of moles of gas present.
(a) I can calculate the number of moles of gas from the ideal gas law. So, I can solve the problem.

We have applied the "four-step approach" to five typical problems in general chemistry, ranging from the simple to the complex. We should caution you that there is nothing magic about this approach; it comes with no guarantees. Each step, particularly (3)—analysis of the principles involved, can require a considerable amount of thought on your part. However, that is precisely what is required to work a difficult problem such as Example 11.5. No amount of number juggling can substitute for a logical analysis based on an understanding of chemical principles.

Checking Answers

After you have worked a problem, you should, if at all possible, check to make sure that your answer makes sense. A mistake, either in reasoning or arithmetic, can lead to an absurd answer to a problem. It is good idea to reflect for a moment on the magnitude of your answer, to see if it seems reasonable. This may save you considerable embarrassment and perhaps even suggest where you have gone wrong. Consider the problem listed in Figure 11.5, where at least two of the four students came up with answers that do not make sense.

Answer (4) is clearly ridiculous; it is impossible to have a negative number of moles. The other nonsensical answer is not quite as easy to spot. However, if you think about it, you will realize that (2) cannot be correct. With only 1.20 mol/ℓ of HI, the maximum concentration of H_2 that could be formed would be 0.60 mol/ℓ. This is

Which are the *incorrect* answers?

Question: The equilibrium constant for the reversible reaction

$$2\ HI(g) \rightleftharpoons H_2(g) + I_2(g)$$

is 0.010. If one starts with a concentration of HI of 1.20 mol/ℓ (no H_2 or I_2 present), what is the concentration of H_2 at equilibrium?

Student Answers: (1) 0.10 mol/ℓ (3) 0.20 mol/ℓ
(2) 1.00 mol/ℓ (4) −0.20 mol/ℓ

FIGURE 11.5

true because 2 mol of HI is required to form 1 mol of H_2, according to the coefficients of the balanced equation. It follows that the answer "1.00 mol/ℓ" for the equilibrium concentration of H_2 is impossible. The two remaining answers, 0.10 mol/ℓ and 0.20 mol/ℓ, are both plausible, since they fall within the limits of 0.00 (no reaction) and 0.60 (complete reaction). As it happens, the correct answer is 0.10 mol/ℓ; the calculation involved is discussed in Example 8.4, p 98.

Ordinarily, to check whether your answer is reasonable, you must have a clear understanding of the terms and principles involved. As a case in point, consider the answer obtained in Example 11.1:

$$\text{mass Na atom} = 3.818 \times 10^{-23}\ g$$

This answer "makes sense" provided you realize that an atom is a tiny, invisible particle, far too small to be weighed on any balance. It should then have an extremely small mass, of the order of magnitude of 10^{-23} g. If, by mistake, you had arrived at a mass of, let us say, 22.99 g for a sodium atom, it should be clear that something is wrong. No atom could possibly weigh that much.

Another illustration of this kind of reasoning occurs in Example 11.5. Here, you should realize that the molar mass *must* be an integral multiple of 29.0 g, (i.e., 29.0 g, 58.0 g, 87.0 g, . . .). This is true because the molecular formula is an integral multiple of the simplest formula. The answer GMM = 58.0 g should be reassuring. The fact that 58.0 is exactly twice 29.0 suggests that your calculation is probably correct. Conversely, if you arrived at a molar mass *less* than 29.0 g (i.e., 10 g, 22 g, . . .), a "warning bell" should go off in your mind. That couldn't possibly be a correct answer. Neither could answers such as 40 g, 70 g, . . . , which are *not* whole-number multiples of 29.0 g.

11.3 COMPLICATIONS; TOO LITTLE OR TOO MUCH INFORMATION

Sometimes, when you analyze a problem in general chemistry, it appears that you do not have enough information to solve it. Conversely, you may have what seems to be excess information, some of it unrelated to the problem at hand. We will consider these two situations separately.

Too Little Data

Certain problems require assumptions which may be a source of concern to you. Consider, for example, the following question:

A sample of gas occupies a volume of 120 cm^3 at 20°C. If the temperature rises to 50°C, what will the new volume be?

Since the volume of a gas depends upon pressure as well as temperature, you may wonder why there is no mention of pressure in the problem. To work the problem, with the information given, you have to assume that the pressure remains constant. It would have been helpful if this condition had been specified. However, instructors (and textbook writers) sometimes fail to do this sort of thing.

Another case of a problem with a hidden assumption is Example 11.2. Can you see what this assumption is? The statement of the problem does not tell us how many moles of NO are available. The analysis we went through assumes, in effect, that there is an excess of NO. If that is the case, then the yield of NO_2 depends only upon the number of moles of O_2 available.

Assumptions such as these seldom bother students in general chemistry. Indeed, in these two cases, they may not even realize there are assumptions hidden in the problems. In contrast, most students know that there is a vital piece of information missing in the following problem.

A sample of hydrogen gas is collected over water at 25°C. The total pressure of the wet gas is 752 mm Hg. Its volume is 19.6 ℓ. Calculate the number of moles of hydrogen in the sample.

If you can't see what is missing here, look back at Example 11.4. Clearly, to solve this problem, you have to obtain the partial pressure of hydrogen. That requires a knowledge of the *vapor pressure of water at 25°C.*

Suppose you were faced with this problem on a homework assignment. What would you do? Logically, you would look for a table giving the vapor pressure of water as a function of temperature. Such a table, which might appear as an appendix in your textbook, would tell you that the vapor pressure of water at 25°C is 24 mm Hg. With this information, the analysis of the problem is straightforward (recall Example 11.4).

Another, somewhat different example:

What is the concentration of OH^- ions in a solution 0.010 M in H^+?

This question is baffling unless you realize that in any water solution, the concentrations of OH^- and H^+ are related by the equation

$$(\text{conc. } OH^-)(\text{conc. } H^+) = 1.0 \times 10^{-14}$$

With this information, you can readily decide that the concentration of OH^- in this case is 1.0×10^{-12} M. The equation just cited is quoted in your textbook. It is used so often that your instructor may expect you to know it and hence not include it on an examination paper.

Sometimes a piece of missing information which appears to be vital at first glance turns out to be unnecessary. How would you answer this question with no other information available?

Write a balanced net ionic equation for the reaction of benzenesulfonic acid, a strong acid, with OH^- ions.

Many students suppose that in order to write this equation, they must know the formula of benzenesulfonic acid. Not so! All you need to do is apply the general principle that the reaction of any strong acid with OH^- ions is represented by the equation:

$$H^+(aq) + OH^-(aq) \rightarrow H_2O$$

In summary, before you give up on a problem because a vital piece of information seems to be missing:

(1) Make sure the information is really essential (remember the benzenesulfonic acid equation).
(2) If you're sure you need the information, consider that it might be
(a) built into a principle you're supposed to know (e.g., the relation between the concentrations of H^+ and OH^-).
(b) available in a table to which you have access (e.g., vapor pressure of water at 25°C).
(3) If you draw a blank in (1) and (2), ask your instructor. He (or the textbook author) may simply have forgotten to supply it.

Too Much Data

Students often worry when they have more information than is needed to work a problem. As a case in point, consider Example 11.6.

Example 11.6 A sample of gas, confined at a constant pressure of 653 mm Hg, occupies a volume of 120 cm³ at 293 K. If the temperature is increased to 323 K, what will the new volume be?

Solution

$$V_2 = V_1 \times \frac{T_2}{T_1} = 120 \text{ cm}^3 \times \frac{323 \text{ K}}{293 \text{ K}} = 132 \text{ cm}^3$$

You will note that it was unnecessary to specify the pressure of the gas, since it remained constant. In that sense, the statement of the problem contained too much data.

The situation described in Example 11.6 is relatively uncommon. More often, difficulties with "excess information" arise in homework assignments. To answer an end-of-chapter problem, you may need to extract a single item from a 20-page chapter (this could be a chemical equation or an equilibrium constant). *Provided you know what you're looking for,* this shouldn't be too difficult. Otherwise, it's virtually impossible.

Often, on the cover page of an examination, you are given a list of items of background information. These may include atomic masses, other constants, and perhaps basic equations or relations. From these, you are expected to select the items you need to work particular problems on the exam. Again, so long as you know what you need, it's relatively easy to separate the wheat from the chaff. Don't be overly concerned if one or more of the items listed don't prove useful. Perhaps they were included by mistake or to trap the unwary.

Sometimes, a question or problem that puzzles you can be answered by stripping away much of the excess verbiage. Indeed, it is often helpful to see if you can reword a problem to make it simpler. As an illustration of this approach, consider Example 11.7.

Example 11.7 When light strikes a photographic film covered with silver bromide, some of the bromide ions are converted to Br atoms, which escape from the film. How many Br atoms would have to be "lost" before one could detect a change in mass, using the most sensitive balance, which weighs to 1.0×10^{-6} g?

Solution

If you think about it for a moment, you will realize that the question raised here is really a very simple one. It can be expressed in a few words:

How many Br atoms are there in 1.0×10^{-6} g?

To answer this question, note that the atomic mass of Br is 79.9. This means that one mole, 6.0×10^{23} atoms, of Br weighs 79.9 g. Hence,

$$\text{no. Br atoms} = 1.0 \times 10^{-6}\ \text{g} \times \frac{6.0 \times 10^{23}\ \text{Br atoms}}{79.9\ \text{g}} = 7.5 \times 10^{15}\ \text{Br atoms}$$

11.4 IF AT FIRST YOU DON'T SUCCEED . . .

We hope that the techniques suggested in this chapter will help you solve most of the problems encountered in general chemistry. However, we may as well be realistic; sooner or later, you will come up against a problem that baffles you. This happens to all of us, from the beginning student to the most experienced instructor. The question is: What should you do when you are stumped by a problem?

Our advice here depends upon whether the stumbling block appears on an examination or a homework assignment. If you can't analyze a problem on an exam, we suggest that you skip over it, at least temporarily. Try to push it into the back of your mind (a neat trick if you can do it). Go on to answer the other questions on the examination. Only when you have done the best you can on the rest of the exam should you return to the problem that bothers you. Quite possibly, while working on other problems, a method of attack will have occurred to you. At any rate, it's a mistake to get so bogged down on one problem that you have to hurry through the rest of the exam, perhaps making careless mistakes.

More frequently, your hang-up will come on a homework assignment, where you are applying a set of chemical principles for the first time. In this case, the first thing you should do is to make an honest effort to work the problem. Don't give up if you don't immediately see how to do it. Think about the problem. If necessary, look up unfamiliar words in the glossary. Decide where in your text the principle required to work the problem is likely to occur. As you reread that section of the text, keep the problem in mind. Look for a principle or an equation that will help you to solve it.

If, at this point, you are still stymied, our advice is simple and obvious: get help! Consult with someone more experienced than you are. This may be a fellow student, a graduate teaching assistant, or your lecturer. There are four suggestions that we would make in this area.

1. TRY TO CARRY YOUR ANALYSIS OF THE PROBLEM AS FAR AS POSSIBLE. In other words, try to find out what it is that bothers you. An explanation will make a great deal more sense if you have thought about the problem beforehand. This applies particularly to problems where more than one principle is involved. Often, you may be able to get halfway to the answer on your own.

2. DON'T BE AFRAID TO ASK QUESTIONS IN CLASS. Many students are reluctant to speak up, perhaps because they don't want to be embarrassed. As a matter of fact, if you can't understand a particular problem, the chances are that most of the class is having trouble with it. Someone has to break the ice, or your instructor won't be able to help you. That's what he's paid for; make sure you get your money's worth.

3. WHEN YOUR INSTRUCTOR IS EXPLAINING HOW TO WORK A PROBLEM, LISTEN TO WHAT HE IS SAYING. Concentrate on the reasoning involved rather than the arithmetic, the analysis of the problem rather than its set-up. Don't waste time writing down the solution. Your goal is to learn how to apply the principles of chemistry to analyze a variety of problems, not to obtain a numerical answer to a specific problem.

4. AFTER GETTING HELP ON A PROBLEM, WORK AN ANALOGOUS ONE, THIS TIME COMPLETELY ON YOUR OWN. Do this as soon as possible, while the principle involved is still fresh in your mind. Analogous problems, involving the same chemical principle, ordinarily can be found in your textbook. If not, ask your instructor to make one up for you, or, better still, devise one of your own.

Students often waste time working the *same* problem over and over again. Once you have seen how to work a problem correctly, there is little to be gained by repeating the process. All you are testing is your memory. To test your understanding of the principle involved, work a different but analogous problem.

PROBLEMS

As in previous chapters, these problems are in matched pairs. However, the system used differs from that followed in Chapters 1–10. Here, the solution is indicated for the problem on the left. That is, the principle involved or the path followed is spelled out. All you have to do is to work out a numerical answer. The matching problem on the right illustrates the same chemical principle. It can be solved following an analysis and path similar to that of the solved problem. Should you have trouble with it, refer to the appropriate section of your text where the principle is discussed.

11.1 A 12.0 g sample of aluminum has a volume of 4.44 cm^3. What is the density of aluminum?

Solution $\text{Density} = \frac{\text{mass}}{\text{volume}}$

$= \_\_\_\_\_\_ \text{ g/cm}^3$

11.2 How many carbon atoms are required to weigh 1.0×10^{-6} g? (atomic mass C = 12.0)

Solution 6.0×10^{23} C atoms weigh 12.0 g, so the number of C atoms can be found by multiplying the mass given by the conversion factor 6.0×10^{23} C atoms/12.0 g. The answer is ______ .

11.3 Find the mass in grams of 1.26 mol of LiOH.

Solution From a table of atomic masses we find that 1 mol of LiOH weighs 6.94 g + 16.00 g + 1.01 g = 23.95 g. To find the mass of 1.26 mol of LiOH, we multiply by the conversion factor 23.95 g/1 mol, obtaining an answer of ______ g.

11.4 Consider the reaction

$$2\ CO(g) + O_2(g) \rightarrow 2\ CO_2(g)$$

How many grams of CO_2 can be formed from 1.60 mol of CO? (atomic mass C = 12.0, O = 16.0).

Solution First, convert moles of CO to moles of CO_2, using the coefficients of the balanced equation

$$2 \text{ mol CO} \triangleq 2 \text{ mol } CO_2$$

In this way, you should find that ______ mol of CO_2 is formed. To convert to grams of CO_2, note that one mole weighs 44.0 g (i.e., 12.0 + 2(16.0) = 44.0). Multiplying the number of moles of CO_2 calculated above by the conversion factor 44.0 g/1 mol gives as an answer ______ g of CO_2.

11.11 The density of mercury is 13.6 g/cm^3. What is the volume, in cubic centimeters, of 12.0 g of mercury?

11.12 What is the mass in grams of 1.0×10^6 carbon atoms?

11.13 With the aid of a table of atomic masses, determine the number of moles in

a. 12.4 g of LiOH *b.* 12.4 g of LiCl

11.14 Using the equation and information in Problem 11.4, calculate the number of grams of CO_2 formed from 1.09 mol of O_2.

11.5 Consider the reaction in Problem 11.4. Calculate the mass in grams of CO_2 that can be obtained from 1.60 g of CO.

Solution A logical three-step path would be to

a. obtain the number of moles of CO using the relation

$$1 \text{ mol CO} = 28.0 \text{ g CO}$$

This gives _____ mol CO.

b. convert moles of CO to moles of CO_2, using the fact that

$$2 \text{ mol CO} \simeq 2 \text{ mol } CO_2$$

This gives _____ mol CO_2.

c. convert moles of CO_2 to grams, using the fact that

$$1 \text{ mol } CO_2 = 44.0 \text{ g}$$

The final answer is _____ g CO_2.

11.15 For the reaction in Problem 11.4, calculate the mass in grams of CO required to produce 1.60 g of CO_2.

11.6 A sample of gas occupies 245 cm^3 at 1.02 atm and 50°C. How many moles of gas are present?

Solution Follow the procedure outlined in Example 11.3 to obtain _____ mol gas.

11.16 What pressure (atmospheres) is exerted by 0.160 mol of a gas occupying a volume of 1.23×10^3 cm^3 at 30°C?

11.7 A sample of oxygen gas is collected over water at 25°C. The total pressure of the wet gas is 726 mm Hg; its volume is 12.4 ℓ. Calculate the number of moles of oxygen in the sample.

Solution Follow the procedure outlined in Example 11.4, step by step, to obtain _____ mol O_2.

11.17 A sample of 0.812 mol of N_2 is collected over water at 25°C. The gas has a volume of 12.2 ℓ. What is

a. the partial pressure of nitrogen, P N_2?
b. the total pressure of the wet gas?

11.8 A certain hydrocarbon has the simplest formula CH. A sample of the gas occupying 0.252 ℓ at 1.12 atm and 137°C weighs 0.655 g. What is the molecular formula of the hydrocarbon?

Solution Follow the procedure of Example 11.5 to obtain _____ for the molecular formula.

11.18 A certain compound contains the two elements nitrogen (atomic mass = 14.0) and oxygen (atomic mass = 16.0). It has a density of 1.40 g/ℓ at 1.00 atm and 127°C. What is the molar mass of the compound? What is its molecular formula?

11.9 Consider the reaction

$$2\ HI(g) \rightarrow H_2(g) + I_2(g)$$

If one starts with 1.40 mol of HI, the number of moles of H_2 formed could *not* be

a. 1.40 *b.* 1.00 *c.* 0.50 *d.* 0.20

Solution See the discussion of Figure 11.5. This eliminates ______ and ______ .

11.10 What is the concentration of H^+ ions in a solution 0.0010 M in OH^-?

Solution Use the equation on p. 162 to calculate the concentration of H^+. ______ M

11.19 Consider the reaction

$$PCl_3(g) + Cl_2(g) \rightarrow PCl_5(g)$$

If one starts with 1.00 mol of PCl_3 and 0.100 mol of Cl_2, the number of moles of PCl_5 formed could *not* be

a. 0.100 *b.* 0.200 *c.* 0.050 *d.* 1.00

11.20 The quantity pH is defined as

$$pH = -\log_{10}(\text{conc. } H^+)$$

What is the pH of a solution in which the concentration of OH^- is 2.0×10^{-5} M?

APPENDIX 1
Symbols Used in Mathematics

$+$	plus
$-$	minus
$\pm$	plus or minus
$\times$	multiplied by; may also be indicated by a dot between the factors or by enclosing the factors within parentheses. $6 \times 2 = 6 \cdot 2 = (6)(2) = 12$
$\div$	divided by; also indicated by writing the divisor under the dividend with a line between, or separating by a slash. $6 \div 2 = \frac{6}{2} = 6/2 = 3$
$=$	equals
$\equiv$	is identical to; is defined as
$\approx$	is approximately
$\backsimeq$	is equivalent to
$>$	is greater than
$<$	is less than
$\leqslant$	is less than or equal to
$\neq$	does not equal
α	is proportional to
$:$	is to; the ratio of
$\therefore$	therefore
$\cdots$	et cetera, e.g., $P = P_1 + P_2 + P_3 + \cdots$
Σ	the sum of, e.g., $P = \Sigma P_i$, means to add all the individual P_i values
∞	infinity
%	per cent
$\bar{x}$	average value of x
σ	standard deviation
!	factorial, e.g., $4! = 1 \times 2 \times 3 \times 4$
$\|x\|$	absolute value of x, without regard to sign
Δx	increment in x. $\Delta x = x_{\text{final}} - x_{\text{initial}}$
e	base of natural logarithms $= 2.718 \cdots$
$\ln x$	natural logarithm of x
$\log x$	base 10 logarithm of x
$\exp x$	e^x; used in complicated expressions such as: $A \exp - \Delta E/RT = Ae^{-\Delta E/RT}$
$f(x)$	function of x
dx	differential of x
dy/dx	derivative of y with respect to x
$\int$	integral
$\int_a^b$	definite integral from a to b

APPENDIX 2

Answers to Problems

CHAPTER 1

Answers are quoted to the appropriate number of significant figures (see Chapter 6).

1.1 a. 7.723 b. −5.253 c. −2 d. 34

1.2 a. 2.208 b. 148 c. −87

1.3 a. 4.02 · · b. 0.710 · · c. −4.02 · · d. 0.710 · ·

1.4 a. −6.628 · · b. 1.0822 · · c. 0.1744 · ·

1.5 a. 1.164 b. −7.41 c. −8.76

1.6 a. 1.214 · · b. −13.8 · · c. 20.9 · ·

1.7 a. 0.583 · · b. 0.211 · · c. 0.254 · ·

1.8 a. 2.62 · · b. 0.239 · · c. 0.0311 · · d. 88.9 · ·

1.9 a. 2.10 · · b. 2.01 · · c. −2.01 · ·

1.10 a. 447 · · b. 1.06 · ·

1.11 a. 0 b. −1 c. +1

1.12 −890.3 kJ

1.13 a. $1.648\ \text{g}/1.235\ \text{cm}^3 = 1.334\ \text{g/cm}^3$

b. $12.6\ \text{cm}^3 \times 2.70\ \text{g/cm}^3 = 34.0\ \text{g}$

c. $\dfrac{12.7\ \text{g}}{2.70\ \text{g/cm}^3} = 4.70\ \text{cm}^3$

1.14 a. $1.60/3.44 = 0.465$ b. $\dfrac{1.60/32.0}{3.44} = 0.0145$

1.15 $\Delta G = -8.10\ \text{kJ} - 300(0.0034)\ \text{kJ} = -9.12\ \text{kJ}$

1.16 $$V = \frac{(1.02\ \text{mol})\left(0.0821\ \dfrac{\ell \cdot \text{atm}}{\text{mol} \cdot \text{K}}\right)(298\ \text{K})}{1.18\ \text{atm}} = 21.1\ell$$

1.17 a. $1.00/2.00 = 0.500$

b. $\dfrac{0.451}{0.451 + 1.324} = 0.254$

1.18 $V = \dfrac{4\pi}{3}(0.262\ \text{nm})^3 = 0.0753\ \text{nm}^3$

1.19 $u = \left(\dfrac{3 \times 8.31 \times 300}{0.0320}\right)^{1/2} = 483\ \text{m/s}$

1.20 a. $K = (0.10)^2(0.10) = 1.0 \times 10^{-3}$

b. $[H_2]^2 = \dfrac{K}{[O_2]} = \dfrac{4.0}{1.5}$; $[H_2] = (4.0/1.5)^{1/2} = 1.6$

CHAPTER 2

2.1 a. $16 \text{ dozen} \times \frac{12 \text{ eggs}}{1 \text{ dozen}} = 192 \text{ eggs}$

b. $594 \text{ apples} \times \frac{1 \text{ dozen}}{12 \text{ apples}} = 49.5 \text{ dozen}$

2.2 a. $40.6 \text{ g} \times \frac{1 \text{ kg}}{10^3 \text{ g}} = 4.06 \times 10^{-2} \text{ kg}$

b. $40.6 \text{ g} \times \frac{2.205 \text{ lb}}{10^3 \text{ g}} = 0.0895 \text{ lb}$

2.3 a. $120 \text{ g} \times \frac{\$1.79}{453.6 \text{ g}} = \$0.474$

b. $120 \text{ g} \times \frac{\$2.12}{453.6 \text{ g}} = \$0.561$

c. $120 \text{ g} \times \frac{\$2.87}{453.6 \text{ g}} = \$0.759$

2.4 a. $5.0 \ \ell \text{ E.G.} \times \frac{3.0 \ \ell \text{ water}}{7.0 \ \ell \text{ E.G.}} = 2.1 \ \ell \text{ water}$

b. $4.0 \ \ell \text{ water} \times \frac{7.0 \ \ell \text{ E.G.}}{3.0 \ \ell \text{ water}} = 9.3 \ \ell \text{ E.G.}$

2.5 a. $29.71 \text{ in. Hg} \times \frac{1 \text{ atm}}{29.92 \text{ in. Hg}} = 0.9930 \text{ atm}$

b. $29.71 \text{ in. Hg} \times \frac{101.3 \text{ kPa}}{29.92 \text{ in. Hg}} = 100.6 \text{ kPa}$

c. $29.71 \text{ in. Hg} \times \frac{760 \text{ mm Hg}}{29.92 \text{ in. Hg}} = 754.7 \text{ mm Hg}$

2.6 a. $1.31 \frac{\text{g}}{\ell} \times \frac{1 \ \ell}{10^3 \text{ cm}^3} = 1.31 \times 10^{-3} \text{ g/cm}^3$

b. $1.31 \frac{\text{g}}{\ell} \times \frac{1 \text{ kg}}{10^3 \text{ g}} \times \frac{10^3 \ \ell}{1 \text{ m}^3} = 1.31 \text{ kg/m}^3$

2.7 $\frac{100 \text{ yd}}{10.0 \text{ s}} \times \frac{60 \text{ s}}{1 \text{ min}} \times \frac{60 \text{ min}}{1 \text{ h}} \times \frac{1 \text{ mile}}{1760 \text{ yd}} \times \frac{1.609 \text{ km}}{1 \text{ mile}} = 32.9 \text{ km/h}$

2.8 a. $7.0 \text{ cup milk} \times \frac{2 \text{ tbsp butter}}{1 \text{ cup milk}} = 14 \text{ tbsp butter}$

b. $2.0 \text{ tbsp flour} \times \frac{1 \text{ cup milk}}{1.5 \text{ tbsp flour}} = 1.3 \text{ cup milk}$

c. $6.0 \text{ tbsp flour} \times \frac{2 \text{ tbsp butter}}{1.5 \text{ tbsp flour}} \times \frac{1 \text{ cup butter}}{16 \text{ tbsp butter}} = 0.50 \text{ cup butter}$

2.9 a. $0.58 \text{ oz} + 1.50 \text{ oz} + 1.50 \text{ oz} = 3.58 \text{ oz}$

b. $6.00 \text{ d} \times \frac{1.50 \text{ oz rum}}{1 \text{ d}} = 9.00 \text{ oz rum}$

c. $6.6 \text{ oz rum} \times \frac{0.58 \text{ oz mix}}{1.50 \text{ oz rum}} = 2.6 \text{ oz mix}$

d. $12 \text{ oz rum} \times \frac{1 \text{ d}}{1.50 \text{ oz rum}} = 8.0 \text{ d}$

2.10 a. $742 \text{ g} \times \frac{314 \text{ kJ}}{10^3 \text{ g}} = 233 \text{ kJ}$

b. $1.00 \text{ kJ} \times \frac{1.00 \text{ kg}}{314 \text{ kJ}} = 3.18 \times 10^{-3} \text{ kg} = 3.18 \text{ g}$

2.11 a. $2.14 \text{ mol} \times \frac{6.022 \times 10^{23} \text{ molecules}}{1 \text{ mol}} = 1.29 \times 10^{24} \text{ molecules}$

b. $3.19 \times 10^{22} \text{ molecules} \times \frac{1 \text{ mol}}{6.022 \times 10^{23} \text{ molecules}} = 0.0530 \text{ mol}$

2.12 a. $4.06 \text{ mol} \times \frac{44.01 \text{ g}}{1 \text{ mol}} = 179 \text{ g}$

b. $4.06 \text{ mol} \times \frac{6.022 \times 10^{23} \text{ molecules}}{1 \text{ mol}} = 2.44 \times 10^{24} \text{ molecules}$

2.13 a. CH_4: $12.01 \text{ g} + 4(1.008 \text{ g}) = 16.04 \text{ g}$
C_2H_2: $2(12.01 \text{ g}) + 2(1.008 \text{ g}) = 26.04 \text{ g}$
C_6H_6: $6(12.01 \text{ g}) + 6(1.008 \text{ g}) = 78.11 \text{ g}$

b. $122 \text{ g } CH_4 \times \frac{1 \text{ mol}}{16.04 \text{ g}} = 7.61 \text{ mol } CH_4$

$122 \text{ g } C_2H_2 \times \frac{1 \text{ mol}}{26.04 \text{ g}} = 4.69 \text{ mol } C_2H_2$

$122 \text{ g } C_6H_6 \times \frac{1 \text{ mol}}{78.11 \text{ g}} = 1.56 \text{ mol } C_6H_6$

2.14 a. $206 \text{ g } AgNO_3 \times \frac{100 \text{ g water}}{952 \text{ g } AgNO_3} = 21.6 \text{ g water}$

b. $21.6 \text{ g water} \times \frac{222 \text{ g } AgNO_3}{100 \text{ g water}} = 48.0 \text{ g } AgNO_3$

2.15 a. $0.0821 \frac{\ell \cdot \text{atm}}{\text{mol} \cdot \text{K}} \times \frac{101.3 \text{ kPa}}{1 \text{ atm}} = 8.32 \frac{\ell \cdot \text{kPa}}{\text{mol} \cdot \text{K}}$

b. $0.0821 \frac{\ell \cdot \text{atm}}{\text{mol} \cdot \text{K}} \times \frac{760 \text{ mm Hg}}{1 \text{ atm}} = 62.4 \frac{\ell \cdot \text{mm Hg}}{\text{mol} \cdot \text{K}}$

c. $0.0821 \frac{\ell \cdot \text{atm}}{\text{mol} \cdot \text{K}} \times \frac{10^3 \text{ cm}^3}{1 \ \ell} = 82.1 \frac{\text{cm}^3 \cdot \text{atm}}{\text{mol} \cdot \text{K}}$

2.16 a. $1.1 \times 10^{-1} \frac{\text{g}}{\ell} \times \frac{1 \ \ell}{10^3 \text{ cm}^3} = 1.1 \times 10^{-4} \text{ g/cm}^3$

b. $1.1 \times 10^{-1} \frac{\text{g}}{\ell} \times \frac{1 \text{ kg}}{10^3 \text{ g}} \times \frac{10^3 \ \ell}{1 \text{ m}^3} = 1.1 \times 10^{-1} \text{ kg/m}^3$

c. $1.1 \times 10^{-1} \frac{\text{g}}{\ell} \times \frac{1 \text{ mol}}{98.1 \text{ g}} = 1.1 \times 10^{-3} \text{ mol}/\ell$

2.17 a. $482 \frac{\text{m}}{\text{s}} \times \frac{1 \text{ km}}{10^3 \text{ m}} \times \frac{3.6 \times 10^3 \text{ s}}{1 \text{ h}} = 1.74 \times 10^3 \text{ km/h}$

b. $482 \frac{\text{m}}{\text{s}} \times \frac{10^2 \text{ cm}}{1 \text{ m}} \times \frac{3.6 \times 10^3 \text{ s}}{1 \text{ h}} \times \frac{24 \text{ h}}{1 \text{ d}} = 4.16 \times 10^9 \text{ cm/d}$

2.18 a. $1.51 \text{ mol } NH_3 \times \frac{5 \text{ mol } O_2}{4 \text{ mol } NH_3} = 1.89 \text{ mol } O_2$

b. $0.282 \text{ mol } NH_3 \times \frac{6 \text{ mol } H_2O}{4 \text{ mol } NH_3} \times \frac{18.0 \text{ g } H_2O}{1 \text{ mol } H_2O} = 7.61 \text{ g } H_2O$

c. $6.40 \text{ g NO} \times \frac{1 \text{ mol NO}}{30.0 \text{ g NO}} \times \frac{4 \text{ mol } NH_3}{4 \text{ mol NO}} = 0.213 \text{ mol } NH_3$

d. $9.80 \text{ g } O_2 \times \frac{1 \text{ mol } O_2}{32.0 \text{ g } O_2} \times \frac{4 \text{ mol NO}}{5 \text{ mol } O_2} \times \frac{30.0 \text{ g NO}}{1 \text{ mol NO}} = 7.35 \text{ g NO}$

2.19 a. $1.61 \text{ mol } Bi^{3+} \times \frac{1 \text{ mol } Bi_2S_3}{2 \text{ mol } Bi^{3+}} = 0.805 \text{ mol } Bi_2S_3$

b. $2.11 \text{ mol } S^{2-} \times \frac{1 \text{ mol } Bi_2S_3}{3 \text{ mol } S^{2-}} \times \frac{514.3 \text{ g } Bi_2S_3}{1 \text{ mol } Bi_2S_3} = 362 \text{ g } Bi_2S_3$

c. $1.00 \text{ g } S^{2-} \times \frac{1 \text{ mol } S^{2-}}{32.1 \text{ g } S^{2-}} \times \frac{2 \text{ mol } Bi^{3+}}{3 \text{ mol } S^{2-}} \times \frac{209.0 \text{ g } Bi^{3+}}{1 \text{ mol } Bi^{3+}} = 4.34 \text{ g } Bi^{3+}$

2.20 a. $\Delta H = 0.682 \text{ mol } CH_4 \times \frac{-890.3 \text{ kJ}}{1 \text{ mol } CH_4} = -607 \text{ kJ}$

b. $-10.0 \text{ kJ} \times \frac{16.04 \text{ g } CH_4}{-890.3 \text{ kJ}} = 0.180 \text{ g } CH_4$

CHAPTER 3

3.1 a. red: $\frac{50}{92} \times 100 = 54$; black 46%

b. 0.54, 0.46

c. total mass = 50(11 g) + 42(13 g) = 1100 g (2 sig. fig.)

Mass % red $= \frac{50(11\text{ g})}{50(11\text{ g}) + 42(13\text{ g})} \times 100 = 50$

3.2 a. $\frac{10}{100} \times 250 = 25$

b. $120 \times \frac{100}{10} = 1200$

3.3 a. $\frac{98}{100} \times 10^3\text{ g} = 980\text{ g}$

b. $\frac{100}{98} \times 10^3\text{ g} = 1020\text{ g}$

3.4 a. nickels: $\frac{50}{70} \times 100 = 71\%$; dimes 29%

b. 50 × 5¢ + 20 × 10¢ = 450¢

c. $\frac{450¢}{70} = 6.4¢$

3.5 a. $10x + 5(80 - x) = 500$

b. $400 + 5x = 500$; $x = 20$; 20 dimes, 60 nickels

c. $\frac{20}{80} \times 100 = 25$; 0.25

3.6 a. 4(40 g) + 3(40 g) + 1(120 g) = 400 g

b. apples: $\frac{160\text{ g}}{400\text{ g}} \times 100 = 40\%$; oranges: $\frac{120\text{ g}}{400\text{ g}} \times 100 = 30\%$; grapefruit: 30%

3.7 a. JB: $\frac{3.0}{5.0} \times 100 = 60\%$; GD: 40%

b. JB: $\frac{60}{100} \times 100\text{ g} = 60\text{ g}$; GD: 40 g

c. JB: $\frac{60\text{ g}}{3.0\text{ g}} = 20$; GD: $\frac{40\text{ g}}{1.0\text{ g}} = 40$

d. GD/JB = 40/20 = 2.0

3.8 a. Cortland/McIntosh = 164/82 = 2.0

Delicious/McIntosh = 205/82 = 2.5

b. McIntosh : Cortland : Delicious = 1.0 : 2.0 : 2.5 = 2 : 4 : 5

3.9 a. 100 − 35 = 65%

b. $\frac{\text{Salt}}{\text{Sand}} = \frac{35}{65} = 0.54$

3.10 a. 2.5 × 20 = 50

b. $\frac{32}{50} \times 100 = 64$

c. theor. yield $= 60 \times \frac{100}{64} = 94$ bushels

94/20 = 4.7 bushels

3.11 a. methyl alcohol: $\frac{1.0}{3.5} \times 100 = 29$; water 71%

b. 0.29, 0.71

c. mass = 1.0(32.0 g) + 2.5(18.0 g) = 77 g

methyl alcohol: $\frac{32.0\text{ g}}{77\text{ g}} \times 100 = 42\%$

3.12 a. $250 \text{ g} \times \frac{69.9}{100} = 175 \text{ g}$

b. $120 \text{ g} \times \frac{100}{69.9} = 172 \text{ g}$

3.13 a. $1.20 \text{ mol} \times 0.7808 = 0.937 \text{ mol}$

b. $1.00 \times \frac{100}{78.08} = 1.28 \text{ mol}$

3.14 $68.93(0.6016) + 70.93(0.3984) = 69.73$

3.15 $64.96(x) + 62.96(1 - x) = 63.54$

$2.00x = 0.58$; $x = 0.29$

29% Cu-65, 71% Cu-63

3.16 a. $40.1 \text{ g} + 2(35.5 \text{ g}) + 6(16.0 \text{ g}) = 207.1 \text{ g}$

b. Ca: $\frac{40.1}{207.1} \times 100 = 19.4\%$; Cl: $\frac{71.0}{207.1} \times 100 = 34.3\%$; O = 46.3%

3.17 a. 19.3 g Na, 26.8 g S, 53.9 g O

b. Na: $19.3 \text{ g} \times \frac{1 \text{ mol}}{23.0 \text{ g}} = 0.839 \text{ mol Na}$

S: $26.8 \text{ g} \times \frac{1 \text{ mol}}{32.1 \text{ g}} = 0.835 \text{ mol S}$

O: $53.9 \text{ g} \times \frac{1 \text{ mol}}{16.0 \text{ g}} = 3.37 \text{ mol O}$

c. 1 S : 1 Na ; 4 O : 1 Na

d. $NaSO_4$

3.18 a. $4.92/1.64 = 3.00$; $5.74/1.64 = 3.50$

b. $C_7H_6O_2$

3.19 a. 80 b. $80/20 = 4.0$

3.20 a. $2.50 \text{ g CO} \times \frac{1.14 \text{ g } CH_3OH}{1 \text{ g CO}} = 2.85 \text{ g } CH_3OH$

b. $\frac{2.24 \text{ g}}{2.85 \text{ g}} \times 100 = 78.6\%$

c. $\frac{2.85 \text{ g}}{0.786} = 3.63 \text{ g } CH_3OH$ theor. yield ; $\frac{3.63 \text{ g}}{1.14} = 3.18 \text{ g CO}$

CHAPTER 4

4.1 a. 1×10^3 b. 1×10^9 c. 1×10^{-6} d. 1.622×10^4
e. 2.126×10^2 f. 1.89×10^{-1} g. 6.18×10^0 h. 7.846×10^{-8}

4.2 a. 0.0012 b. 64 c. 3.0 d. 410,000 e. 0.000 001 4

4.3 a. 3×10^3 b. 10,000 c. 2×10^{-4} d. 4×10^8
e. 1.5×10^{-2} f. same

4.4 $2.0 \times 10^{-5} < 7.1 \times 10^{-5} < 6.0 \times 10^{-2} < 3.6 \times 10^{-1} < 4.6 \times 10^3$

4.5 a. 9.30×10^{12} b. 6.3×10^5 c. 1.05×10^{-2} d. 3.90×10^{16}
e. 1.26×10^3

4.6 a. 5.18×10^{-3} b. 2.0×10^1 c. 3.38×10^{15} d. 1.48×10^{-18}

4.7 a. 4.67×10^{-6} b. 1.2×10^{14} c. 3.6×10^{-41} d. 3.0×10^3
e. 9.2×10^2

4.8 a. 2.5×10^{-1} b. 4.02×10^{-4} c. 2.78×10^3 d. 2.0×10^6
e. 3.1×10^1

4.9 a. 4.71×10^4 b. 6.20×10^4 c. 8.05×10^5 d. 9.75×10^4

4.10 a. 5.47×10^{-2} b. 2.45×10^{-4} c. 6.37×10^{-10} d. 6.02×10^{23}

4.11 mass: 6.65×10^{-24} g radius: 4.6×10^{-9} cm speed: 1.36×10^5 cm/s

4.12 0.000 013 ; 1.6×10^{-2} ; 4.5×10^{-2}
4.13 B is fastest, C is slowest.
4.14 $Fe(OH)_3 < Al(OH)_3 < Cr(OH)_3 < Fe(OH)_2 < Mg(OH)_2$
4.15 He: 4.003 g (lightest) ; N: 14.00 g ; Sr: 87.62 g (heaviest)
4.16

conc. H^+	conc. OH^-	conc. H^+/conc.OH^-
4.0×10^{-11}	2.5×10^{-4}	1.6×10^{-7}
3.6×10^{-8}	2.8×10^{-7}	1.3×10^{-1}
10^{-6}	10^{-8}	10^2

4.17

n	1	2	3
r	5.3×10^{-9}	2.1×10^{-8}	4.7×10^{-8}

4.18

1.0	1.8×10^{-4}	1.8×10^{-4}
0.10	5.7×10^{-5}	5.7×10^{-4}
0.010	1.8×10^{-5}	1.8×10^{-3}

4.19 $p\ H_2 = 1.224 \times 10^3$ mm Hg $- 9.80 \times 10^2$ mm Hg $= 2.44 \times 10^2$ mm Hg
4.20 12.02 kJ + 0.65 kJ = 12.67 kJ

CHAPTER 5

5.1 a. 0.912 b. 0.2180 c. −2.3082 d. −6.00 e. −9.35
5.2 a. 7.640 b. 5.53×10^1 c. 6.90×10^{-4} d. 1.21×10^{-7} e. 10^{-126}
5.3 a. 886 b. 7.07×10^1 c. -1.6×10^{-3}
5.4 a. 1.3×10^2 b. 7.6×10^1 c. 2.1
5.5 a. −3.80 b. −1.90 c. −7.60
5.6 a. $\log x - \log y$ b. $\log x + \log y$ c. $\log x + 2 \log y$
5.7 $\log x = 2 \log y - \frac{1}{2} \log z$
5.8 a. $y = 12.4 - 2.0 \log 3^{8/5} = 12.4 - 3.2(\log 3) = 10.9$

b. $\log x^{8/5} = \dfrac{-11.4}{-2.0}$; $\log x = \dfrac{11.4}{3.2}$; $x = 3.7 \times 10^3$

5.9 a. 1.796 b. 2.000 c. −9.69
5.10 a. 2.72 b. 8.21×10^{-2} c. 4.084×10^5
5.11 a. 6.00 b. 1.583 c. 8.52 d. −0.78
5.12 a. 1×10^{-4} b. 2.5×10^{-13} c. 7.2×10^{-4} d. 10
5.13 a. $\Delta G^0 = -2.303(8.314)(300)\log(0.020) = 9.8 \times 10^3 J$

b. $\log K = \dfrac{-1.08 \times 10^4}{(2.303)(8.314)(400)} = -1.41$; $K = 3.9 \times 10^{-2}$

5.14 a. $\log \dfrac{1.0}{0.10} = \dfrac{0.045\, t}{2.303} = 1.0$; $t = 51$ min

b. $\log \dfrac{2.00}{x} = \dfrac{(2.0 \times 10^{-3})50}{2.303} = 0.043$; $\dfrac{2.00}{x} = 1.1$; $x = 1.8$

5.15 conc. $H^+ = 1.0 \times 10^{-4}$
conc. $S^{2-} = (1 \times 10^{-21})/(1 \times 10^{-8}) = 1 \times 10^{-13}$
5.16 Take the logarithms of both sides:
$\log$ (conc. H^+) + $\log$(conc. OH^-) = −14.00
−pH − pOH = −14.00; pH + pOH = 14.00
5.17 Take the logarithm of both sides:
$\log K_a = 2 \log$(conc. H^+) − $\log$(conc. HAc)
$\log K_a = -2$pH − $\log$(conc. HAc)
5.18 a. $E = +0.75\ V + 1.6(0.0591) \log$(conc. H^+)
$= +0.75\ V - 0.0946$ pH
b. $+0.47\ V$ c. 2.6

5.19 a. $\Delta G^0 = -(8.314)(250) \ln 16 = -5.8 \times 10^3$

b. $\ln K = \dfrac{1.00 \times 10^4}{(8.314)(100)} = 12.0$; $K = 1.6 \times 10^5$

5.20 $f = 1.76 \times 10^{-2}$

CHAPTER 6

6.1 a. 4 b. 3 c. 4 d. 3 e. 5

6.2 a. 3 b. 3 c. 2 d. 1 e. 2 or 3

6.3 a. 9.49×10^{-3} b. 1.2×10^2 c. 1.2 d. 1.68

6.4 a. 1.02 b. 1.3×10^4 c. 2.4 d. 5.80×10^1

6.5 2.3×10^{-3}

6.6 a. 1.0×10^{-5} b. 1.77×10^{-3}

6.7 a. 15.33 g b. 5.77 cm^3 c. 33.7 cm d. 15.25 m

6.8 a. 1.06×10^2 g b. 3.34×10^{-1} cm c. 75.2 cm^2 d. 0.52 g/cm^2

6.9 a. 0.2047 b. 0.72 c. 3.338 d. 1×10^2 e. 1.53 f. 1.6×10^3

6.10 a. 0.4158 b. −7.59

6.11 a. 4 b. 4 c. 3 d. 3 e. 4 f. 2 g. 2

6.12 4, 3, 4, exact number, 2

6.13 $V = \dfrac{9.9654 \text{ g}}{0.9970 \text{ g/cm}^3} = 9.995 \text{ cm}^3$

6.14 a. 7.328 g b. 23 g

6.15 a. 222 g/mol

6.16 a. $(4.09 \times 10^{-8} \text{ cm})^3 = 6.84 \times 10^{-23} \text{ cm}^3$

b. $6.84 \times 10^{-23} \text{ cm}^3 \times 6.022 \times 10^{23} = 41.2 \text{ cm}^3$

6.17 617 g

6.18 a. 2.040 g − 0.721 g − 0.050 g = 1.269 g

b. $\% \text{ C} = \dfrac{0.721}{2.040} \times 100 = 35.3$; $\% \text{ H} = \dfrac{0.050}{2.040} \times 100 = 2.5\%$

$\% \text{ I} = \dfrac{1.269}{2.040} \times 100 = 62.21\%$

6.19 a. 10^{-4} b. 8×10^{-5} c. 7.6×10^{-5} d. 7.62×10^{-5}

6.20 $\Delta H = 2.303\, RT(B - \log P) = 2.319 \times 10^4$ J

CHAPTER 7

7.1 a. $x = 1.5 \times 10^{-4}$ b. $x = 1$ c. 4 ; −16

7.2 a. 6.0 b. 1.6×10^7 c. 3.8×10^6

7.3 a. $x = \dfrac{10.0\, y}{6 + 180\, y}$ b. 0.0538, 0.0417

7.4 a. $x = 1.5/30\, yz$ b. 0.05 ; 0.005

7.5 a. $x = uv/yz$ b. $u = xyz/v$ c. $z = uv/xy$

7.6 $z = y/x = uv/wx$

7.7 a.
$$\begin{array}{r} 6x - 8y = 32 \\ \underline{6x + 15y = 78} \\ -23y = -46 \end{array} ; \quad y = 2,\ x = 8$$

b.
$$\begin{array}{r} 0.0556x + 0.0556y = 0.01112 \\ \underline{0.0556x + 0.0153y = 0.00709} \\ 0.0403y = 0.00403 \end{array} ; \quad y = 0.100,\ x = 0.100$$

7.8 $\quad x + 2y = 3z$

$\quad \underline{3z + 3x = 18y}$

$\quad 4x + 2y = 18y\,; \qquad 4x = 16y\,; \qquad x = 4y$

7.9 $\quad 4x + 2y = 12z$

$\quad 3x - 9y = 12z\,;$

hence, $4x + 2y = 3x - 9y$

7.10 $\quad 2x + y = u + 2y$

$\quad u + y = 2x$

7.11 a. $T = 167$ b. $t = 46$ c. $0.600T = 273\,;\ T = 455$

7.12 a. $V = \dfrac{(1.00\ \text{mol})(0.0821\ \ell \cdot \text{atm/mol} \cdot \text{K})(300\ \text{K})}{0.952\ \text{atm}} = 25.9\ \ell$

b. $T = \dfrac{(1.961\ \text{atm})(29.6\ \ell)}{(0.500\ \text{mol})(0.0821\ \ell \cdot \text{atm/mol} \cdot \text{K})} = 1.41 \times 10^3\ \text{K}$

7.13 a. $\text{m} = \dfrac{c}{d - c\text{M}_2/1000}$

b. $\text{m} = \dfrac{0.1000}{1.004 - 0.1000 \times 184/1000} = \dfrac{0.1000}{0.986} = 0.1014\ \text{m}$

7.14 $\text{M}_2 = \dfrac{1.86 \times \text{g}_2 \times 1000}{-\text{t}_\text{f} \times \text{g}_1} = \dfrac{1.86 \times 1.00 \times 1000}{0.423 \times 20.0} = 220$

7.15 a. $V_2 = V_1 \times \dfrac{P_1}{P_2} \times \dfrac{T_2}{T_1}$ b. $V_1 = V_2 \times \dfrac{P_2}{P_1} \times \dfrac{T_1}{T_2}$ c. $P_2 = P_1 \times \dfrac{V_1}{V_2} \times \dfrac{T_2}{T_1}$

d. $P_1 = P_2 \times \dfrac{V_2}{V_1} \times \dfrac{T_1}{T_2}$ e. $T_1 = T_2 \times \dfrac{V_1}{V_2} \times \dfrac{P_1}{P_2}$

7.16 $d = \dfrac{g}{V} = \dfrac{P\ (\text{GMM})}{RT}$

7.17 a. $x + y = 1$

b. $36.97x + 36.97y = 36.97$

$\quad \underline{36.97x + 34.97y = 35.45}$

$\quad 2.00y = 1.52\,; \qquad y = 0.760,\ x = 0.240$

7.18

$$\begin{array}{r} 4\ \text{Zn(s)} \rightarrow 4\ \text{Zn}^{2+}\ (\text{aq}) + 8\ e^- \\ \underline{\text{NO}_3^-\ (\text{aq}) + 10\ \text{H}^+\ (\text{aq}) + 8\ e^- \rightarrow \text{NH}_4^+\ (\text{aq}) + 3\ \text{H}_2\text{O}} \\ 4\ \text{Zn(s)} + \text{NO}_3^-\ (\text{aq}) + 10\ \text{H}^+\ (\text{aq}) \rightarrow 4\ \text{Zn}^{2+}\ (\text{aq}) + \text{NH}_4^+\ (\text{aq}) + 3\ \text{H}_2\text{O} \end{array}$$

7.19

$$\begin{array}{ll} 2\ \text{H}_2\text{O}_2(\text{l}) \rightarrow 2\ \text{H}_2\text{O}(\text{l}) + \text{O}_2(\text{g})\,; & \Delta H = +196.4\ \text{kJ} \\ \underline{2\ \text{H}_2\text{O}(\text{l}) \rightarrow 2\ \text{H}_2(\text{g}) + \text{O}_2(\text{g})\,;} & \underline{\Delta H = +571.6\ \text{kJ}} \\ 2\ \text{H}_2\text{O}_2(\text{l}) \rightarrow 2\text{H}_2(\text{g}) + 2\ \text{O}_2(\text{g})\,; & \Delta H = +768.0\ \text{kJ} \end{array}$$

7.20

$$\begin{array}{ll} \text{N}_2(\text{g}) + 2\ \text{O}_2(\text{g}) \rightarrow 2\ \text{NO}_2(\text{g})\,; & \Delta H = +69.8\ \text{kJ} \\ \underline{2\ \text{NO}_2(\text{g}) \rightarrow 2\ \text{NO}(\text{g}) + \text{O}_2(\text{g})\,;} & \underline{\Delta H = +113.0\ \text{kJ}} \\ \text{N}_2(\text{g}) + \text{O}_2(\text{g}) \rightarrow 2\ \text{NO}(\text{g})\,; & \Delta H = +182.8\ \text{kJ} \end{array}$$

CHAPTER 8

8.1 a. 2 ; 3 b. n ; $3n/2$ c. 4 ; $7 - 3n/2$

8.2

52	$-x$	$52 - x$
94	$-2x$	$94 - 2x$
12	$+x$	$12 + x$

8.3

90 kg	−10 kg	80 kg	90	$-x/6$	$90 - x/6$
70 kg	−10 kg	60 kg	70	$-x/6$	$70 - x/6$
500 kg	−40 kg	460 kg	500	$-2x/3$	$500 - 2x/3$
0 kg	+60 kg	60 kg	0	$+x$	x

8.4 a. $x = \pm 4.5 \times 10^{-3}$
b. $x/(1 - x) = \pm 2.4$; $x = 0.71$ or 1.7
c. $x/(0.10 - 2x) = \pm 1.0$; $x = 0.033$ or 0.10

8.5 a. $x/(1 - x) = 18^{1/3} = 2.6$; $x = 0.72$
b. $(2 - x)/(2 + 2x) = \pm 4.0$; $x = -0.67$ or -1.4
c. $1 - x = \pm 0.25x$; $x = 0.80$ or 1.3

8.6 a. $x = 0.270$ or -0.370
b. $x = 0.0950$ or -0.105
c. $x = 0.0311$ or -0.0321

8.7 a. $x = 1.5$ or -6.5
b. $x = 2.4$ or -12.4
c. $x = 0.019$ or -0.53

8.8 a. $x = 0.42$ or -0.72
b. $x = 0.365$ or 0.685

8.9 a. $x = \pm 0.316$ (17, 15%)
b. $x = \pm 0.100$ (5%)
c. $x = \pm 0.0316$ (1.6%)

8.10 $x^2 \approx 2.0 \times 10^{-2}$; $x = 0.14$
$x^2 = 0.86(2.0 \times 10^{-2})$; $x = 0.13$

8.11 a. x b. $0.40 - 2x$

8.12 a. x b. $1.00 - x$ c. $0.100 - x$

8.13

2.04	−0.50	1.54	2.04	$-x$	$2.04 - x$
4.62	−1.50	3.12	4.62	$-3x$	$4.62 - 3x$
0.00	+1.00	1.00	0.00	$+2x$	$2x$

8.14 a. $\dfrac{[H_2] \times [I_2]}{[HI]^2} = 0.015$

b. $\dfrac{x^2}{(0.40 - 2x)^2} = 0.015$; $\dfrac{x}{0.40 - 2x} = 0.12$; $x = 0.039$

8.15 a.

1.00	$-x$	$1.00 - x$
1.00	$-x$	$1.00 - x$
0.00	$+2x$	$2x$

b. $\dfrac{(1.00 - x)^2}{4x^2} = 0.015$; $\dfrac{1.00 - x}{2x} = 0.12$; $x = 0.81$

8.16 a. $7.00 \times 10^{-4} = \dfrac{[H^+] \times [F^-]}{[HF]}$

b. $7.00 \times 10^{-4} = \dfrac{x^2}{1.00 - x}$; $x = 2.61 \times 10^{-2}$

8.17 a.

0.00	$+x$	x
0.50	$+x$	$0.50 + x$
1.00	$-2x$	$1.00 - 2x$

b. $\dfrac{x(0.50 + x)}{(1.00 - 2x)^2} = 0.015$; $0.94x^2 + 0.56x - 0.015 = 0$; $x = 0.026$

8.18

	orig. conc.	change	equil. conc.
PCl_5	1.00	$-x$	$1.00 - x = 0.64$
PCl_3	0.00	$+x$	$x = 0.36$
Cl_2	0.00	$+x$	$x = 0.36$

$0.20 = \dfrac{x^2}{1.00 - x}$; $x = 0.36$

8.19 $x^2 = 7.00 \times 10^{-4}$; $x = 2.65 \times 10^{-2}$; 1.5% ; yes

8.20 $\dfrac{x^2}{(2.00 - x)} = 1.0 \times 10^{-2}$; $x^2 \approx 2.0 \times 10^{-2}$; $x \approx 0.14$

$\dfrac{x^2}{1.86} = 1.0 \times 10^{-2}$; $x^2 = 1.86 \times 10^{-2}$; $x = 0.14$

CHAPTER 9

9.1 a. 7/2 b. 7 c. 49/2 d. −35/2 e. 7/4

9.2 a. $y = 5x$; 20 b. $y = 2x^2$; 32 c. $y = 3x$; 11/3 d. 9

9.3 a. inverse b. direct c. inverse d. neither

9.4 a. $a(x_2 - x_1)$ b. $a\left[\dfrac{1}{x_2} - \dfrac{1}{x_1}\right]$ c. $a(x_2 - x_1)$ d. $a(x_2^{1/2} - x_1^{1/2})$

e. $a\left[\dfrac{1}{x_2^{1/2}} - \dfrac{1}{x_1^{1/2}}\right]$

9.5 a. $y_2/y_1 = x_1^{1/2}/x_2^{1/2} = (0.50)^{1/2} = 0.71$

b. $x_2^{1/2} = x_1^{1/2}y_1/y_2$; $x_2 = x_1 y_1^2/y_2^2 = 1.0/4.0 = 0.25$

9.6 a. $16.0 = -61.5 - 8.0k$; $k = -77.5/8.0 = -9.7$

b. $y = -61.5 + 9.7(4.0) = -22.7$

c. $0 = -61.5 + 9.7x$; $x = 6.3$

9.7 a. $\log(y_2/y_1) = a(x_1 - x_2)$ b. $y_2 - y_1 = a \log(x_1/x_2)$

9.8 a. $\log y = 0.00$; $y = 1.0$ b. $\log y = 1.00$; $y = 10$

c. $0.477 = \dfrac{-2.00}{x} + 2.00$; $1.52x = 2.00$; $x = 1.32$

9.9 a. $a = 24$

b. $6 = 4a + b$; $b = 2$; $a = 1$

c. $a = 1$; $1 = b + c$; $0 = 2b + 4c$; $b = 2$, $c = -1$

9.10 a. $3.6 = 3.0a + b$; $6.0 = 5.0a + b$; $a = 1.2$, $b = 0$

b. $0.5 = 1.0a + b$; $3.0 = 2.0a + b$; $a = 2.5$, $b = -2.0$

c. $6.0 = b$; $3.5 = a + 6.0$; $a = -2.5$

9.11 0, 40, 80, 120, 200

9.12 a. $\dfrac{\text{rate}_2}{\text{rate}_1} = \left(\dfrac{\text{conc.}_2}{\text{conc.}_1}\right)^n$; $2 = 2^n$, $n = 1$

b. $4 = 2^n$, $n = 2$ c. $1 = 2^n$, $n = 0$

9.13 $2.00 = \left(\dfrac{0.447}{0.316}\right)^n = 1.41^n$; $n = 2$

$1.00 = K(0.316)^2$; $K = 10.0$

$[F_2] = 10.0[F]^2$

9.14 a. $u = aT^{1/2}/M^{1/2}$ b. $\dfrac{u_2}{u_1} = \dfrac{T_2^{1/2}}{T_1^{1/2}} \times \dfrac{M_1^{1/2}}{M_2^{1/2}}$

9.15 a. rate $= k/M^{1/2}$; $\dfrac{\text{rate } CH_4}{\text{rate He}} = \left(\dfrac{4.00}{16.0}\right)^{1/2} = 0.500$

b. $0.82 = \left(\dfrac{32.0}{M}\right)^{1/2}$; $32.0/M = 0.67$; $M = 48$

9.16 a. $\Delta S = \dfrac{\Delta H - \Delta G}{T} = \dfrac{-198.2 \text{ kJ} + 140.0 \text{ kJ}}{300 \text{ K}} = \dfrac{-58.2 \text{ kJ}}{300 \text{ K}} = -0.194 \text{ kJ/K}$

b. $\Delta G = -198.2 \text{ kJ} + 500(0.194)\text{kJ} = -101.2 \text{ kJ}$

c. $T = \dfrac{\Delta H}{\Delta S} = \dfrac{-198.2 \text{ kJ}}{-0.194 \text{ kJ/K}} = 1.02 \times 10^3\text{K}$

9.17 $\Delta G^0_2 - \Delta G^0_1 = (2.303)(8.314)T \log (K_1/K_2)$

9.18 a. $\log \frac{760}{P_1} = \frac{4.70 \times 10^4(23)}{(2.303)(8.314)(373)(350)} = 0.432$

$760/P_1 = 2.71$; $P_1 = 280$ mm Hg

b. $\log \frac{1000}{760} = 0.119 = \frac{(4.70 \times 10^4)(T - 373)}{(2.303)(8.314)(373)T}$

$(T - 373)/T = 0.0181$; $T = 373/0.9819 = 380$ K

9.19 $\frac{[NO_2]^2}{[NO]^2 \times [O_2]}$; 9.99, 9.99, 9.99, 10.0 ; $K = 9.99$

9.20 $0.074089 = 40.00a + b$

$0.073688 = 10.00a + b$

$a = 0.000401/30.00$; $a = 1.34 \times 10^{-5}$

$b = 0.074089 - 40.00a = 0.073554$

CHAPTER 10

10.1 See graph.

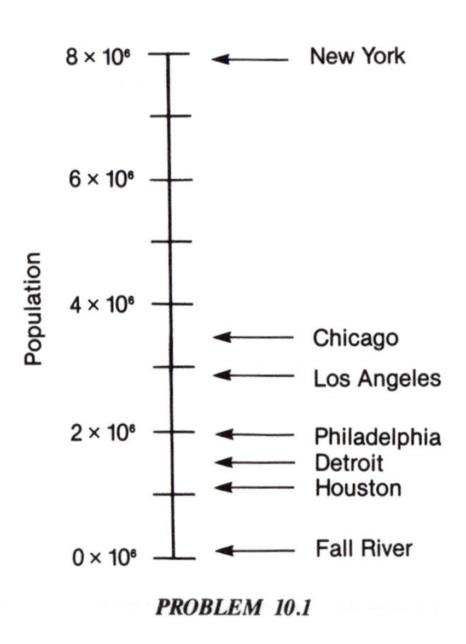

PROBLEM 10.1

10.2 a. $x = 5, y = 8$; $x = 6, y = 9.5$ b. 5, 7.2 c. 2.4, 4.4

10.3 a. 86°, 162° b. 38°, 93°

10.4 a. 6.0 b. 10.5 c. 0.70

10.5 See graphs (pp. 181–182).

10.6 $b = 3.0$; $a = 10$

10.7 See graphs (pp. 183–185). c) is a straight line with $y = 4.0/x^2$

10.8 a. $y = 2x^2$ (see graph, p. 186) b. $y = 2/z$ (see graph, p. 187)

10.9 a.

$\log y$	0.000	0.500	0.667	0.750
$1/x$	1.00	0.500	0.333	0.250

b. See graph (p. 188).

c. slope $= -1.00 = -A$; $A = 1.00$

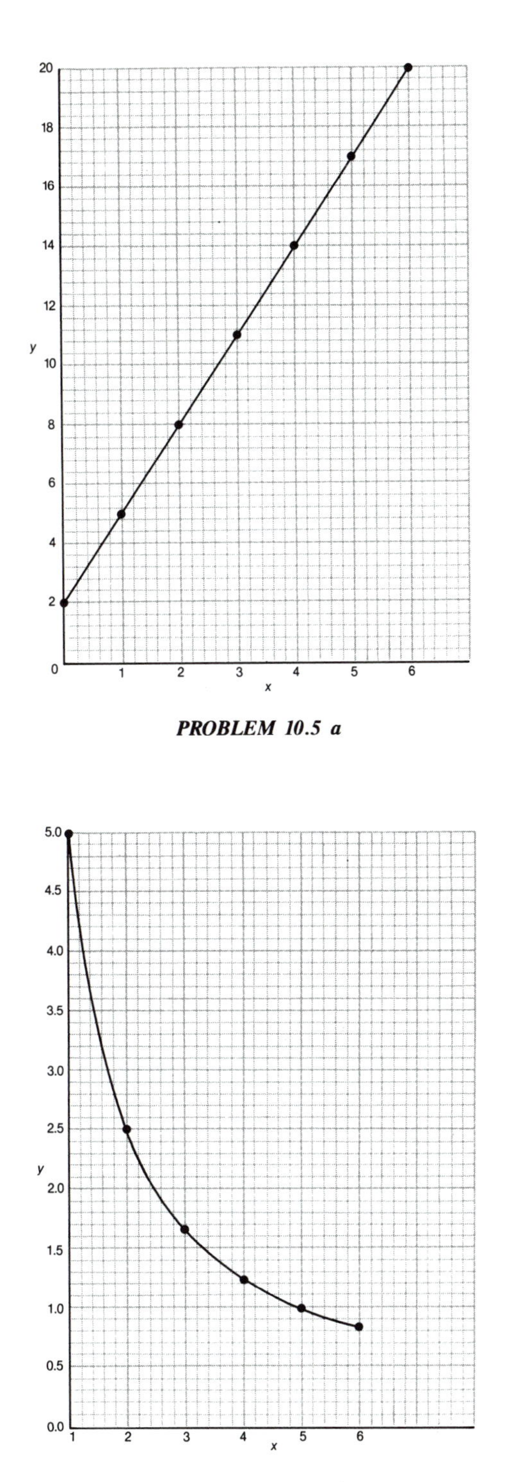

PROBLEM 10.5 a

PROBLEM 10.5 b

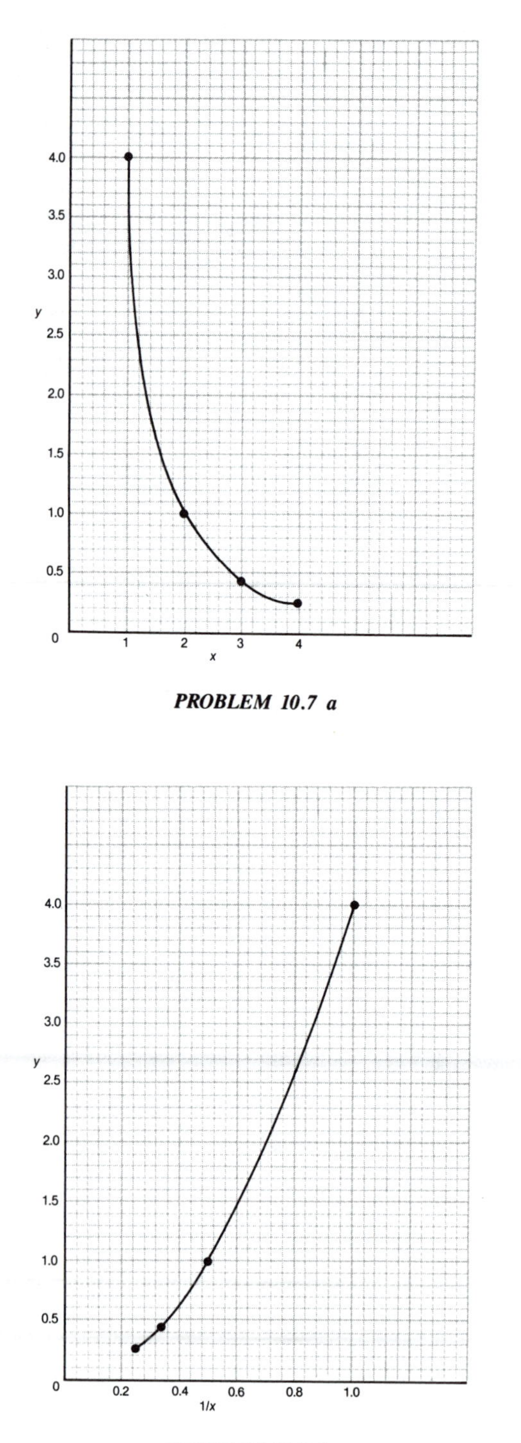

PROBLEM 10.7 a

PROBLEM 10.7 b

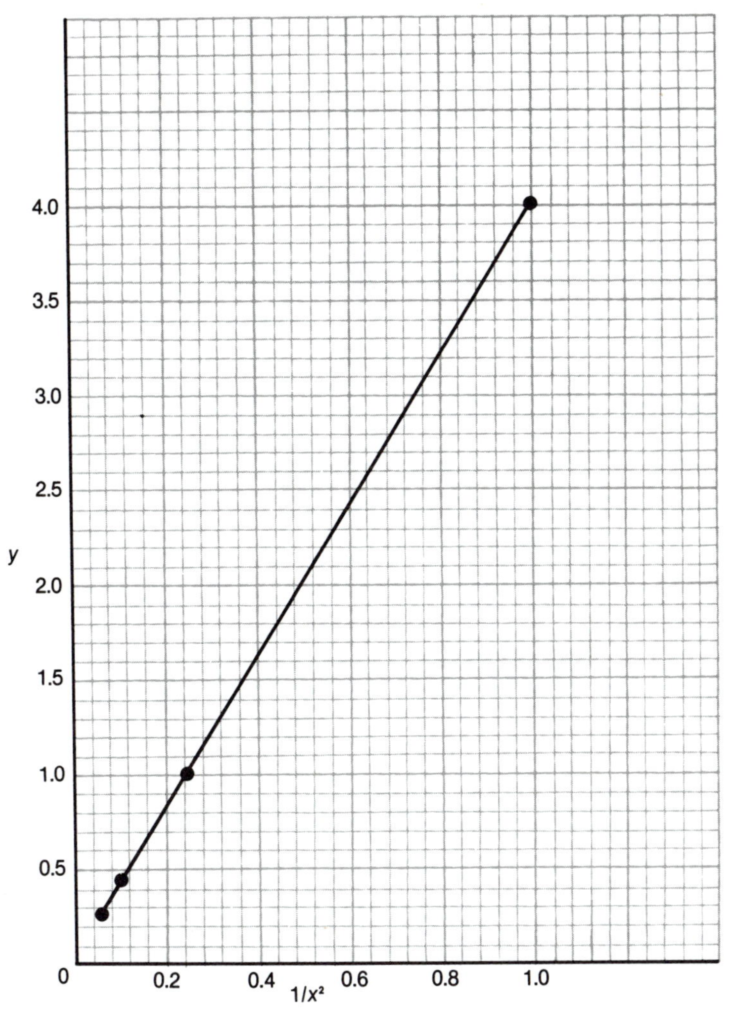

PROBLEM 10.7 c

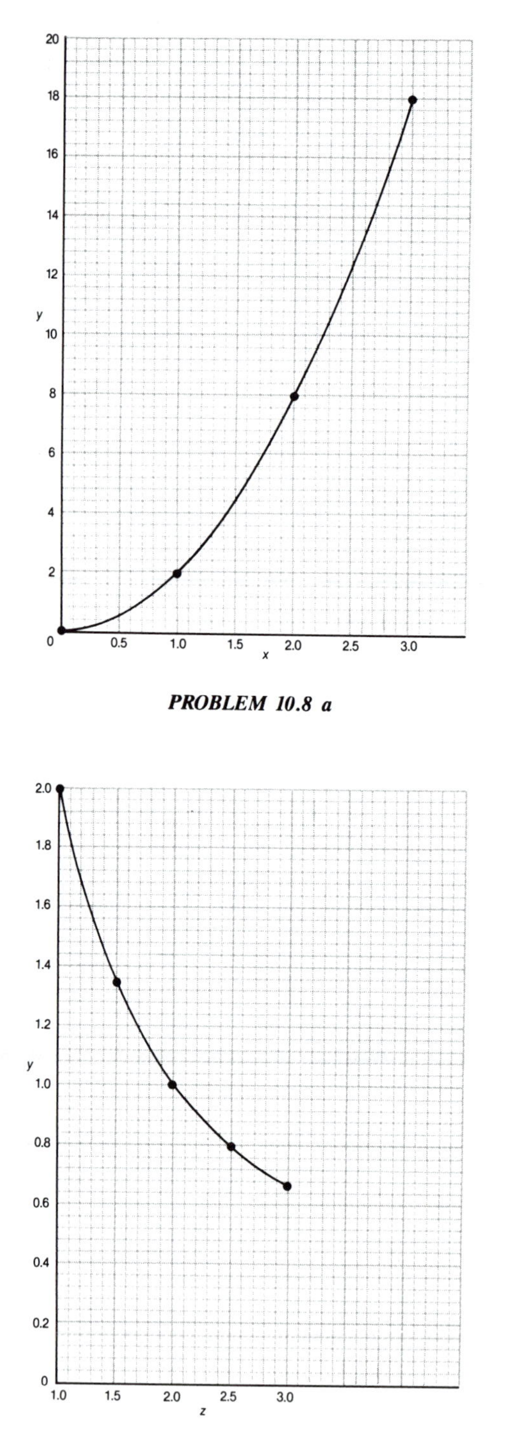

PROBLEM 10.8 a

PROBLEM 10.8 b

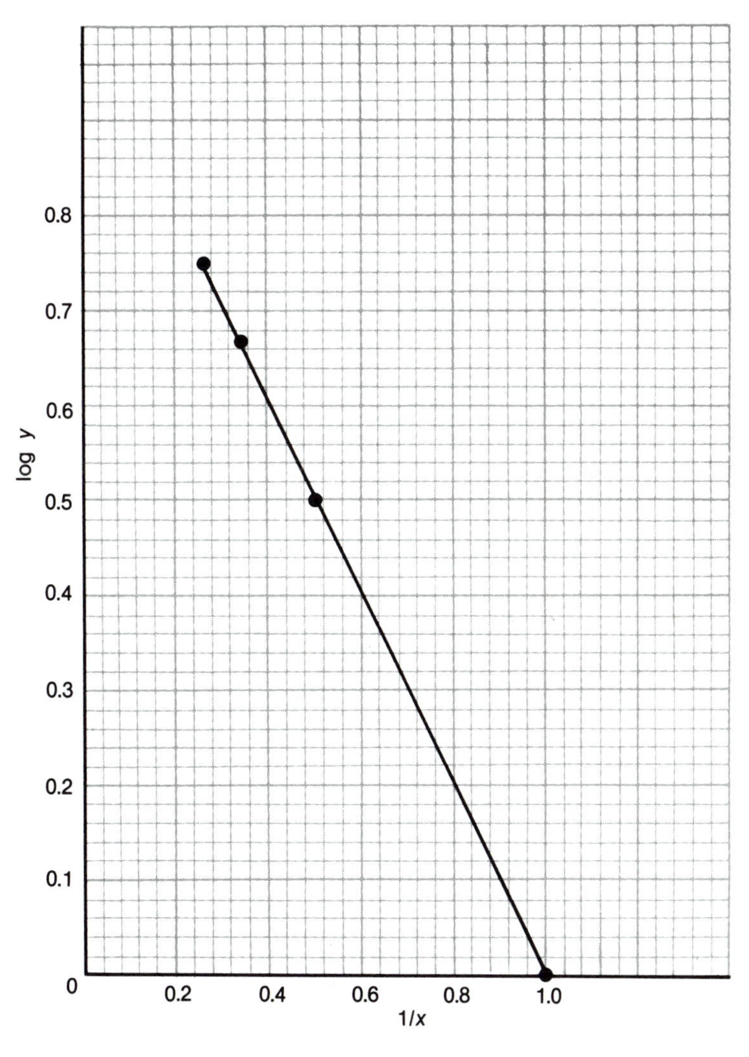

PROBLEM 10.9 b

10.10

y	x	yx	x^2
−1.2	0	0.0	0
1.5	1	1.5	1
4.2	2	8.4	4
7.0	3	21.0	9
9.6	4	38.4	16
12.5	5	62.5	25
33.6	15	131.8	55

$33.6 = 6b + 15a$

$131.8 = 15b + 55a$

$a = 2.7 \qquad b = -1.2$

10.11 See graph.

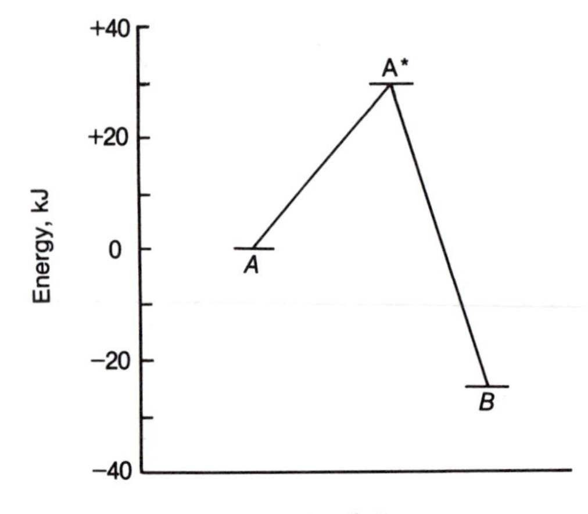

PROBLEM 10.11

10.12 a. 21 g/100 g water b. 57°C c. 81°C

10.13 a. +6.8 kJ b. +1.5 kJ c. 230 K

10.14 a. −22.8 kJ b. −52.7 kJ c. 465 K

10.15 See graph (p. 190).

10.16 $\Delta H = y$ intercept $= -92.4$ kJ

$\Delta S = -\text{slope} = -0.199$ kJ/K

10.17 a. See graph, p. 188 b. See graph, p. 188. (straight line).

c. conc. $CrO_4^{2-} = \dfrac{a}{(\text{conc. } Ag^+)^2}$; a = slope in (b) $= 1.0 \times 10^{-12}$

conc. $CrO_4^{2-} \times (\text{conc. } Ag^+)^2 = 1.0 \times 10^{-12}$

10.18 a. See graph, p. 193 ; $P = 8.21 \times 10^{-3}\ T$

b. See graph, p. 194 ; $PV = 24.6\ \ell \cdot \text{atm}$

10.19 a.

$\log P$	1.785	2.002	2.203	2.391
$1/T$	3.663×10^{-3}	3.534×10^{-3}	3.413×10^{-3}	3.300×10^{-3}

b. See graph, p. 190.

c. slope $= 0.606/(-0.363 \times 10^{-3}) = -1.67 \times 10^3$

$\Delta H_{vap} = (1.67 \times 10^3)(2.303)(8.314) = 3.20 \times 10^4$ kJ

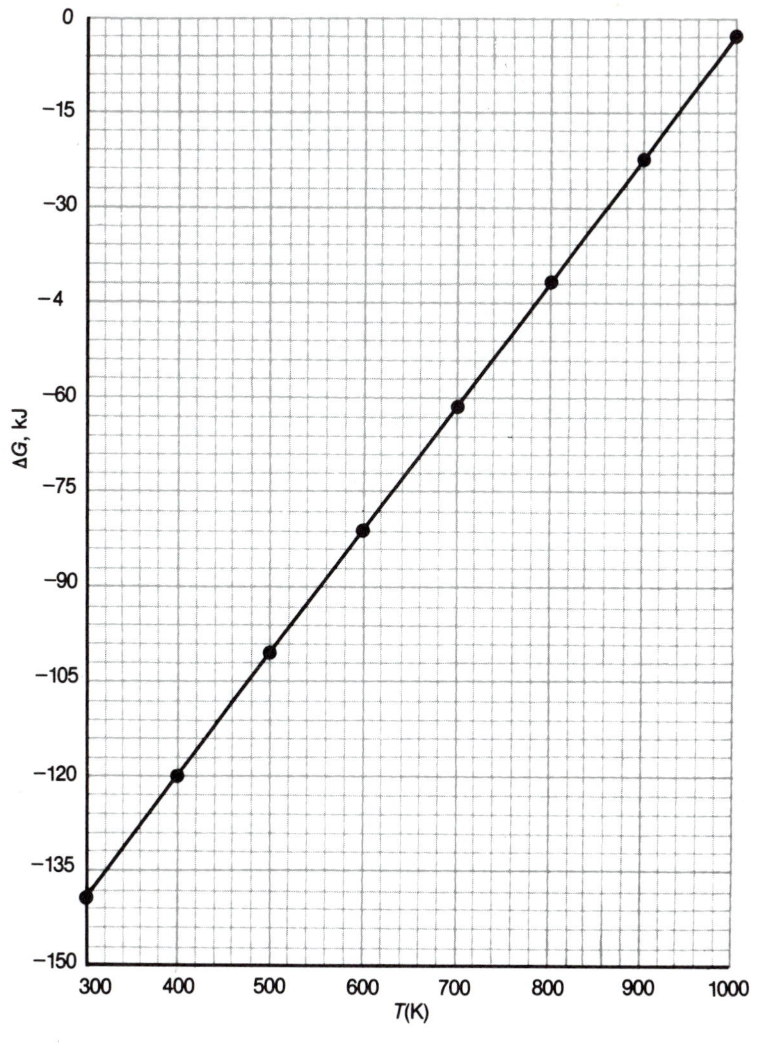

PROBLEM 10.15

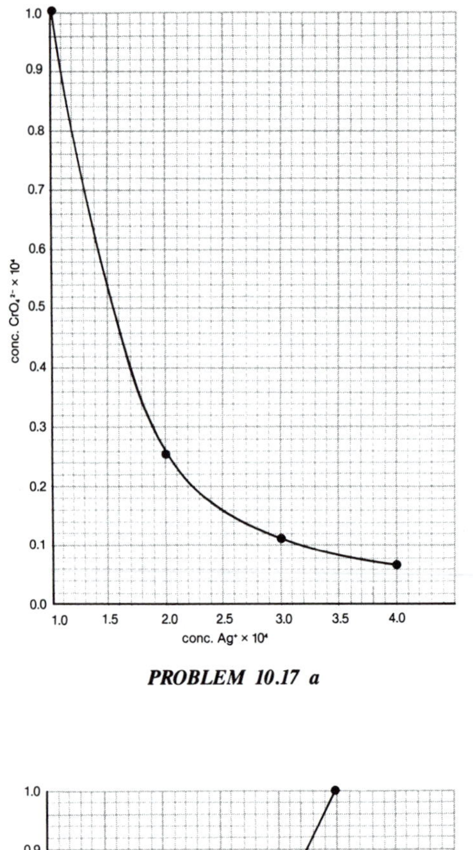

PROBLEM 10.17 a

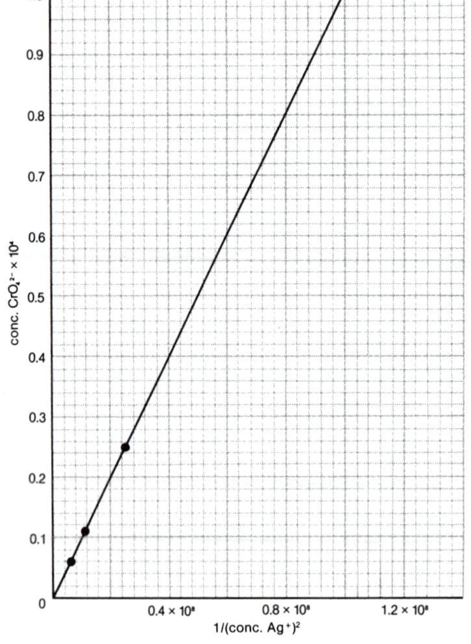

PROBLEM 10.17 b

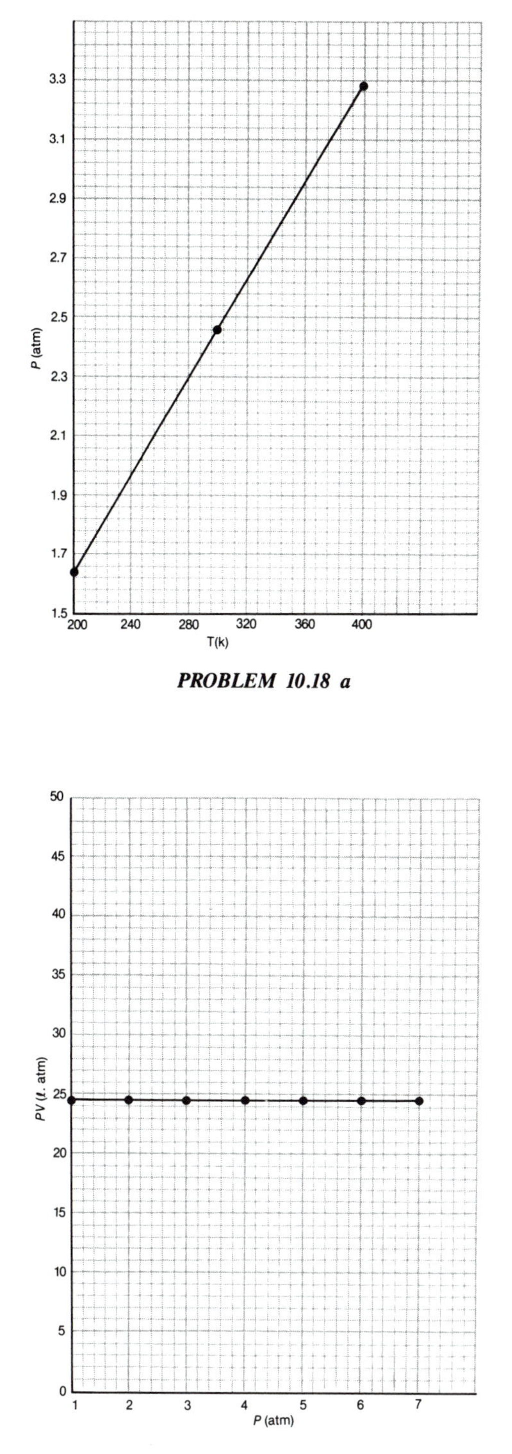

PROBLEM 10.18 a

PROBLEM 10.18 b

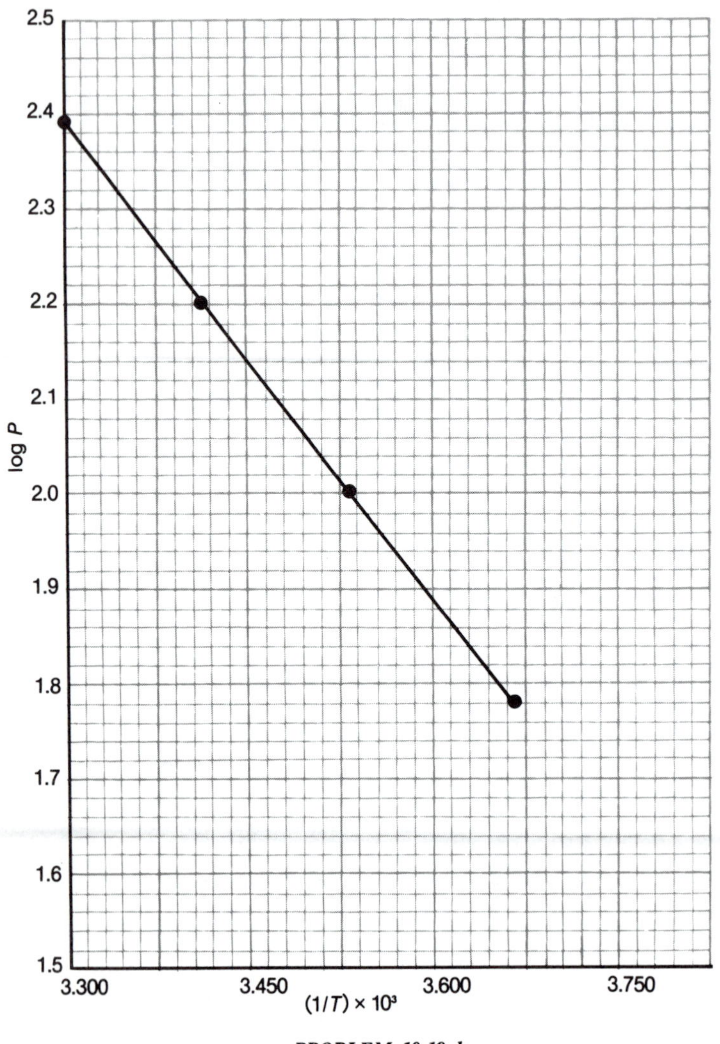

PROBLEM 10.19 b

10.20

y	x	yx	x^2
2.8	1.0×10^2	2.8×10^2	1.0×10^4
4.8	2.0×10^2	9.6×10^2	4.0×10^4
6.0	3.0×10^2	18.0×10^2	9.0×10^4
7.7	4.0×10^2	30.8×10^2	16.0×10^4
9.4	5.0×10^2	47.0×10^2	25.0×10^4
30.7	15.0×10^2	108.2×10^2	55.0×10^4

$30.7 = 5b + 1500a$

$1.082 \times 10^4 = 1500b + 550\,000a$

$a = 0.0161$; $b = 1.31$

$\Delta H = +1.3$ kJ ; $\Delta S = -0.016$ kJ/K

CHAPTER 11

11.1 2.70 g/cm^3

11.2 5.0×10^{16}

11.3 30.2 g

11.4 70.4 g

11.5 a. 0.0571 mol CO b. 0.0571 mol CO_2 c. 2.51 g CO_2

11.6 9.42×10^{-3} mol

11.7 0.468 mol O_2

11.8 MM = 78.1 ; C_6H_6

11.9 *a, b*

11.10 1.0×10^{-11} M

11.11 V = mass/density = 0.882 cm^3

11.12 $1.0 \times 10^6 \text{ C atoms} \times \dfrac{12.0 \text{ g}}{6.0 \times 10^{23} \text{ C atoms}} = 2.0 \times 10^{-17} \text{ g}$

11.13 a. $12.4 \text{ g} \times \dfrac{1 \text{ mol}}{23.95 \text{ g}} = 0.518 \text{ mol}$ b. $12.4 \text{ g} \times \dfrac{1 \text{ mol}}{42.39 \text{ g}} = 0.293 \text{ mol}$

11.14 $1.09 \text{ mol } O_2 \times \dfrac{2 \text{ mol } CO_2}{1 \text{ mol } O_2} \times \dfrac{44.0 \text{ g } CO_2}{1 \text{ mol } CO_2} = 95.9 \text{ g } CO_2$

11.15 $1.60 \text{ g } CO_2 = \dfrac{1 \text{ mol } CO_2}{44.0 \text{ g } CO_2} \times \dfrac{1 \text{ mol CO}}{1 \text{ mol } CO_2} \times \dfrac{28.0 \text{ g CO}}{1 \text{ mol CO}} = 1.02 \text{ g CO}$

11.16 $P = \dfrac{nRT}{V} = 3.24$ atm

11.17 a. $P = \dfrac{nRT}{V} = 1.63$ atm

b. $P\ H_2O = \dfrac{24}{760} \text{ atm} = 0.032$ atm ; $P_{tot} = 1.66$ atm

11.18 molar mass = 46.0 g ; NO_2

11.19 *b, d*

11.20 conc. $H^+ = 5.0 \times 10^{-10}$ M ; pH = 9.30

Subject Index

Note; Page numbers in italics refer to illustrations; *t* indicates tables.

Chemical Topics Index

(p = page, E = Example, P = Problem)

A

B

C

D

E

F

G

H

I

S

T

U

V

W

Y